职业院校智能制造专业"十三五"系列教材

智能控制技术专业英语

主　编　张明文　王璐欢
副主编　霰学会　宁　金
参　编　顾三鸿　王　伟
主　审　高文婷

机械工业出版社

本书针对智能控制技术专业应用人才培养的需要而编写，以培养和提高读者的智能控制技术专业英语能力为目标，旨在使读者掌握智能控制技术专业的英语知识和基础理论。本书由15个单元组成，内容涵盖了智能控制领域的主要技术分支，包括智能制造、机电一体化技术、工业控制技术、工业机器人技术等。

本书图文并茂，通俗易懂，实用性强，既可作为智能控制技术、机械电子、电气自动化及工业机器人技术等相关专业的英语教材，也可供从事相关工作的技术人员参考使用。

This book is written to meet the needs of the training of application talents in the field of intelligent control technology. It aims to cultivate and improve the English proficiency of intelligent control technology for readers, and to enable readers to master the English knowledge and basic theory of intelligent control technology. This book consists of 15 units, covering the main technical branches in the field of intelligent control, including intelligent manufacturing, mechatronics technology, industrial control technology, industrial robot technology, etc.

This book is rich in illustrations, easy to understand and practical. It can be used not only as a professional English textbook for intelligent control technology, mechanical electronics, electrical automation and industrial robot technology, but also as a reference for technical personnel engaged in related fields.

图书在版编目（CIP）数据

智能控制技术专业英语/张明文，王璐欢主编．—北京：机械工业出版社，2020.12（2024.3重印）

职业院校智能制造专业"十三五"系列教材

ISBN 978-7-111-67253-1

Ⅰ.①智… Ⅱ.①张…②王… Ⅲ.①智能控制-英语-高等职业教育-教材
Ⅳ.①TP273

中国版本图书馆 CIP 数据核字（2021）第 002967 号

机械工业出版社（北京市百万庄大街 22 号　邮政编码 100037）
策划编辑：王振国　责任编辑：王振国
责任校对：陈玉芝　责任印制：李　昂
北京中科印刷有限公司印刷
2024 年 3 月第 1 版第 4 次印刷
184mm×260mm・14.25 印张・348 千字
标准书号：ISBN 978-7-111-67253-1
定价：49.80 元

电话服务　　　　　　　　　　网络服务
客服电话：010-88361066　　机　工　官　网：www.cmpbook.com
　　　　　010-88379833　　机　工　官　博：weibo.com/cmp1952
　　　　　010-68326294　　金　　书　　网：www.golden-book.com
封底无防伪标均为盗版　　　机工教育服务网：www.cmpedu.com

Preface

With the rapid development of the new generation of information technology and manufacturing technology, intelligent manufacturing technology has become an objective trend of the development of the world's manufacturing industry, and the world's major industrialized countries are vigorously promoting and applying it. At present, with the rise of labor cost, the gradual disappearance of demographic dividend, the transformation of production mode to flexible, intelligent and refined direction, it is urgent to build a new intelligent manufacturing system. The development of intelligent manufacturing is not only in line with the internal requirements of the development of China's manufacturing industry, but also an inevitable choice to reshape the new advantages and realize transformation and upgrading of China's manufacturing industry. "Made in China 2025" proposes to take intelligent manufacturing as the main direction of deep integration of industrialization and information technology, and raises the intelligent manufacturing industry to the level of national strategy.

Intelligent control technology is an important part of intelligent manufacturing technology, which refers to the network and information integration of industrial robots, traditional automation equipment and advanced manufacturing equipment. Intelligent control technology is the combination of mechanical and electronic engineering technology and control technology, including the design and simulation of intelligent system, intelligent system maintenance, system operation, test analysis and management.

In the critical stage of the transformation and upgrading of traditional manufacturing industry, more and more enterprises will face the awkward situation of "easy to get equipment, difficult to find talents". Therefore, to achieve intelligent manufacturing, talent cultivation should go first. According to the guidance for talent development planning of manufacturing industry jointly issued by the Ministry of Education, the Ministry of Human Resources and Social Security and the Ministry of Industry and Information Technology, we should vigorously cultivate talents in need of technical skills in ten key areas of manufacturing industry, and strengthen the construction of teaching resources and basic platform for vocational skills training. In order to better promote the application of intelligent control technology, it is necessary to write a comprehensive introduction textbook of intelligent control technology.

This book is written to meet the needs of the training of application talents in the field of intelligent control technology. It aims to cultivate and improve the English proficiency of intelligent control technology for readers, and to enable readers to master the English knowledge and basic theory of intelligent control technology. This book consists of 15 units, covering the main technical branches in the field of intelligent control, including intelligent manufacturing, mechatronics technology, industrial control technology, industrial robot technology, etc.

This book is rich in illustrations, easy to understand and practical. It can be used not only as a professional English textbook for intelligent control technology, mechanical electronics, electrical automation and industrial robot technology, but also as a reference for technical personnel engaged in related fields. In order to improve the teaching effect, it is suggested to adopt heuristic teaching, open learning and group discussion; in the learning process, it is suggested to combine the supporting teaching auxiliary resources of this book, such as teaching courseware, video materials, teaching reference and expansion materials, etc.

Due to the limited level and time of the editor, there may be some mistakes in the book. Please comment and correct. Any comments and suggestions are welcome and can be fed back to e-mail: market@ jijiezhi. com.

Editor

前 言

随着新一代信息技术与制造技术的飞速发展,智能制造技术已成为世界制造业发展的客观趋势,世界上主要工业发达国家正在大力推广和应用。当前,随着我国劳动力成本上升、人口红利的逐渐消失,生产方式向柔性化、智能化、精细化方向转变,构建新型智能制造体系迫在眉睫。发展智能制造既符合我国制造业发展的内在要求,也是重塑我国制造业新优势、实现转型升级的必然选择。《中国制造2025》提出将智能制造作为工业化、信息化深度融合的主攻方向,将智能制造产业上升到国家战略层面。

智能控制技术是智能制造技术的重要组成部分,是指对工业机器人、传统自动化设备、先进制造装备进行网络化、信息化集成。智能控制技术是机械电子工程技术与控制技术相结合的产物,具体内容包括对智能系统的设计与仿真,智能系统维护、系统运行、试验分析与管理。

在传统制造业转型升级的关键阶段,越来越多的企业将面临"设备易得、人才难求"的尴尬局面。因此,要实现智能制造,人才培养要先行。教育部、人力资源和社会保障部、工业和信息化部编制的《制造业人才发展规划指南》指出,要面向制造业十大重点领域大力培养技术技能紧缺人才,加强职业技能培训教学资源和基础平台建设。针对这一现状,为了更好地推广智能控制技术的应用,亟须编写一本系统、全面的智能控制技术的入门教材。

本书针对智能控制技术专业应用人才培养的需要而编写,以培养和提高读者的智能控制技术专业英语能力为目标,旨在使读者掌握智能控制技术专业的英语知识和基础理论。本书由15个单元组成,内容涵盖了智能控制领域的主要技术分支,包括智能制造、机电一体化技术、工业控制技术、工业机器人技术等。

本书图文并茂,通俗易懂,实用性强,既可作为智能控制技术、机械电子、电气自动化及工业机器人技术等相关专业的专业英语教材,也可供从事相关工作的技术人员参考使用。为了提高教学效果,在教学方法上,建议采用启发式教学,开放性学习,重视小组讨论;在学习过程中,建议结合本书配套的教学辅助资源,如教学课件、视频素材、教学参考与拓展资料等。

由于编者水平及时间有限,书中难免存在不足之处,敬请读者批评指正。任何意见和建议可反馈至 E-mail:market@jijiezhi.com。

编者

CONTENTS 目录

Preface
前言

Unit 1 National Strategy in the Background of Industrial Revolution 工业变革背景下的国家战略 ·················· 1
    Part 1   Industry 4.0 工业4.0 ·················· 1
    Part 2   American National Strategy of Advanced Manufacturing Industry 美国先进制造业国家战略 ·················· 4
    Part 3   Intelligent Manufacturing in China 中国智能制造 ·················· 7
    Part 4   Other National Strategies 其他国家战略 ·················· 10

Unit 2 Theme of Intelligent Manufacturing 智能制造的主题 ·················· 14
    Part 1   Intelligent Factory 智能工厂 ·················· 14
    Part 2   Intelligent Production 智能生产 ·················· 18
    Part 3   Intelligent Logistics 智慧物流 ·················· 21

Unit 3 Key Technology of Intelligent Manufacturing 智能制造关键技术 ·················· 24
    Part 1   Artificial Intelligence 人工智能 ·················· 24
    Part 2   Industrial Internet of Things 工业物联网 ·················· 26
    Part 3   Internet of Things and Cyber-Physical System 物联网与网络物理系统 ·················· 29
    Part 4   Other Technologies 其他技术 ·················· 31

Unit 4 Mechanical Elements and Mechanisms 机械元件和机构 ·················· 35
    Part 1   Kinematic Sketch of Mechanism 机构运动简图 ·················· 35
    Part 2   Mechanical Transmission Mechanism 机械传动机构 ·················· 38
    Part 3   Mechanical Connection Components 机械连接部件 ·················· 42
    Part 4   Manufacturing Process 制造工艺 ·················· 48

Unit 5 Electrical and Electronic Technology 电工电子技术 ·················· 52
    Part 1   Electrical Foundation 电工基础 ·················· 52
    Part 2   Electronic Components 电子元器件 ·················· 55
    Part 3   Kirchhoff's Law 基尔霍夫定律 ·················· 58
    Part 4   Integrated Circuit 集成电路 ·················· 60

Unit 6 Control Theory 控制理论 ·················· 64

Contents
目 录

    Part 1   Composition of the Control System 控制系统构成 ································ 64
    Part 2   Open-Loop and Closed-Loop Control 开环和闭环控制 ························· 66
    Part 3   PID Control PID 控制 ·· 70
    Part 4   Intelligent Control 智能控制 ·· 72

Unit 7   Sensors and Measurements Technology 传感检测技术 ································ 77
    Part 1   Resistive Sensor and Capacitive Sensor 电阻式传感器和电容式传感器 ········· 77
    Part 2   Inductive Sensor 电感式传感器 ·· 80
    Part 3   Temperature Sensor 温度传感器 ··· 84
    Part 4   Photoelectric Sensor 光电式传感器 ·· 87
    Part 5   Optical Grating Sensor 光栅传感器 ·· 90
    Part 6   Radio-Frequency identification 射频识别（RFID） ··························· 92

Unit 8   Hydraulics and Pneumatics Transmission 液压与气压传动 ···························· 96
    Part 1   Hydraulic Transmission System 液压传动系统 ·································· 96
    Part 2   Features and Application of Hydraulic System 液压系统的特点及应用 ········ 100
    Part 3   Pneumatic Transmission System 气压传动系统 ································ 103
    Part 4   Application and Features of Pneumatic System 气压系统的特点及应用 ······· 106

Unit 9   Motor Drive 电动机拖动 ··· 111
    Part 1   DC Motor 直流电动机 ··· 111
    Part 2   Induction Motor 交流异步电动机 ··· 117
    Part 3   Stepper Motor 步进电动机 ·· 122
    Part 4   Servo Motor 伺服电动机 ·· 125

Unit 10   Microcomputer and Microprocessor 微机和微处理器 ································ 129
    Part 1   Microprocessor Technology 微处理器技术 ····································· 129
    Part 2   Single Chip Microcontroller 单片机 ·· 133
    Part 3   Programmable Logic Device 可编程逻辑器件 ·································· 136
    Part 4   Artificial Intelligence（AI）Chip 人工智能芯片 ······························· 139

Unit 11   Industrial Control Core 工业控制核心 ·· 142
    Part 1   Programmable Logic Controller 可编程序控制器 ······························ 142
    Part 2   Motion Controller 运动控制器 ·· 147
    Part 3   Human-Machine Interface 人机界面 ··· 151

Unit 12   Industrial Robot 工业机器人 ·· 155
    Part 1   About Industrial Robot 关于工业机器人 ··· 156
    Part 2   Component and Types of Industrial Robot 工业机器人的组成和分类 ········· 162
    Part 3   Technical Parameters of Industrial Robot 工业机器人的技术参数 ············ 169

Unit 13   Industry Application of Robot 机器人行业应用 ······································ 174
    Part 1   Handling Robot 搬运机器人 ·· 174

Part 2	Welding Robot 焊接机器人	176
Part 3	Painting Robot 喷涂机器人	179
Part 4	Polishing Robot 打磨机器人	182

Unit 14　Machine Vision Technology 机器视觉技术　187
Part 1	Introduction to Machine Vision 机器视觉概述	187
Part 2	Technical Basis of Machine Vision 机器视觉技术基础	190
Part 3	Vision System 工业机器人视觉系统	196
Part 4	Industry Application of Machine Vision 机器视觉工业应用	199

Unit 15　Intelligent Robot 智能机器人　202
Part 1	Definition and Classification of Intelligent Robot 智能机器人的定义与分类	202
Part 2	Basic Elements of Intelligent Robots 智能机器人的基本要素	207
Part 3	Application Analysis of Intelligent Robots 智能机器人应用分析	210
Part 4	Development Trend of Intelligent Robots 智能机器人发展趋势	212

Unit 1 National Strategy in the Background of Industrial Revolution

Part 1 Industry 4.0
工业 4.0

1. What is Industry 4.0?

Industry 4.0 is a name given to the current trend of automation and data exchange in manufacturing technologies. It includes cyber-physical systems, the Internet of Things, cloud computing and cognitive computing. Industry 4.0 is commonly referred to as the fourth industrial revolution. The term "Industry 4.0", shortened to I4.0 or simply I4, originates from a project in the high-tech strategy of the German government, which promotes the computerization of manufacturing.

Video 1

The characteristics given for the German government's Industry 4.0 strategy are: the strong customization of products under the conditions of highly flexible production. The required automation technology is improved by the introduction of methods of self-optimization, self-configuration, self-diagnosis, cognition and intelligent support of workers in their increasingly complex work.

2. The Development of the Four Industrial Revolutions

The development of the four industrial revolutions is shown in Figure 1-1.

Industry 1.0: The first industrial revolution was brought about by mechanization and the harnessing of steam power. This was a big change in industry as labor went from being solely manual to machine-based.

Industry 2.0: The second industrial revolution came with the implementation of electricity in manufacturing. This revolution led to mass production and tremendous economic growth in the U.S..

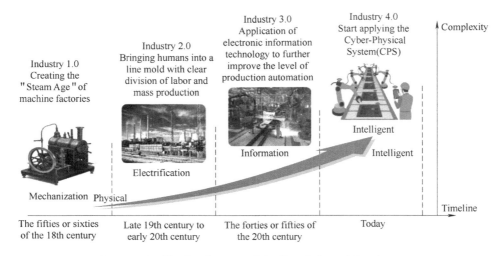

Figure 1-1　The Development of the Four Industrial Revolutions

Industry 3.0: The third revolution in industry is characterized by the introduction of IT technologies to advance automation and motion control in manufacturing.

Industry 4.0: The fourth industrial revolution is the computerization of manufacturing. Industry 4.0 is putting cyber-physical systems to use in order to unlock the potential of industry and manufacturing.

3. Impact of Industry 4.0

Industry 4.0 will affect many areas, most notably:

1) Reliability and continuous productivity.
2) Security issues of Internet of things.
3) Manufacturing sales.
4) Workers'education and skills.

Vocabulary　词汇

1. cognitive　　　　['kɒgnɪtɪv]　　　　adj. 认知的，认识的
2. revolution　　　　[revə'luːʃn]　　　　n. 革命；旋转；运行；循环
3. strategy　　　　['strætədʒɪ]　　　　n. 战略，策略
4. implementation　　[ˌɪmplɪmen'teɪʃn]　　n. 实现；安装启用
5. customization　　　['kʌstəmaɪzeɪʃən]　　n. 定制；[计] 用户化；客制化服务
6. flexible　　　　['fleksəbl]　　　　adj. 灵活的；柔韧的；易弯曲的
7. tremendous　　　　[trɪ'mendəs]　　　　adj. 极大的，巨大的；惊人的；极好的
8. automation　　　　[ˌɔːtə'meɪʃn]　　　　n. 自动化
9. computerization　　[kəmˌpjuːtəraɪ'zeɪʃn]　n. （电子）计算机化（工作）

Notes　注释

1. Industry 4.0　　　　　　　　　　　　　　工业4.0

Unit 1 National Strategy in the Background of Industrial Revolution
工业变革背景下的国家战略

2. cyber-physical system　　　网络物理系统
3. Internet of things　　　物联网
4. cloud computing　　　云计算
5. cognitive computing　　　认知计算
6. data exchange　　　数据交换
7. manufacturing technology　　　制造技术
8. flexible production　　　柔性生产
9. self-optimization　　　自我优化
10. self-configuration　　　自我配置
11. self-diagnosis　　　自我诊断
12. steam power　　　蒸汽动力
13. IT technology　　　信息技术

Reference Translation　参考译文

1. 什么是工业 4.0？

工业 4.0 是当前制造技术自动化和数据交换趋势下的一个名称。它包括网络物理系统、物联网、云计算和认知计算。工业 4.0 通常被称为第四次工业革命。"工业 4.0" 一词的缩写为 I4.0 或简称 I4，源于德国政府高科技战略中的一个项目，用于促进制造业的计算机化。

德国政府工业 4.0 战略的特点是：在高度灵活生产的条件下，产品的强定制。通过引入工人日益复杂的工作中的自我优化、自我配置、自我诊断、认知和智能支持等方法，提高了所需的自动化技术。

2. 四次工业革命的发展历程

四次工业革命的发展历程如图 1-1 所示。

工业 1.0：第一次工业革命是由机械化和蒸汽动力带来的。这是工业界一个很大的变化，因为劳动力从单纯的体力劳动变成了机器化。

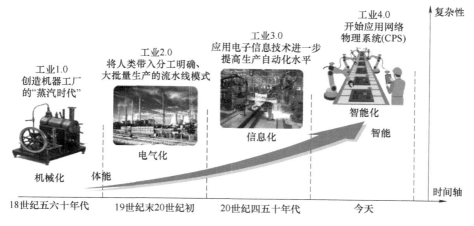

图 1-1　工业革命的发展历程

工业 2.0：第二次工业革命是随着电力在制造业中的实施而发生的。这场革命导致了美国的大规模生产和巨大的经济增长。

工业 3.0：第三次工业革命的特点是引入了信息技术，以推进制造业的自动化和运动控制。

工业 4.0：第四次工业革命，是制造业的计算机化。工业 4.0 正在使用网络物理系统，以释放工业和制造业的潜力。

3. 工业 4.0 的影响

"工业 4.0"将影响许多领域，最显著的是：

1. 可靠性和持续的生产率。
2. 物联网安全问题。
3. 生产销售。
4. 工人的教育培养和技能。

Part 2 American National Strategy of Advanced Manufacturing Industry
美国先进制造业国家战略

1. American Manufacturing and Competitiveness

American Manufacturing plays a vital role in the U.S. economy. Advances in manufacturing enable the economy to continuously improve as new technologies and innovations increase productivity, enable new products, and create entirely new industries. These new industries often create new jobs that replace low-skilled jobs.

Video 2

2. National Strategy of Advanced Manufacturing Industry

In October 2014, the United States released the "Revitalization of Advanced Manufacturing Industry in the United States" report. The report is used to guide the federal government's plans and actions to support the research and development in advanced manufacturing. The contents of three pillars are shown in Table 1-1.

Table 1-1　Contents of Three Pillars

Name	Contents
Accelerating technological innovation	To formulate the national advanced manufacturing strategy, increase the priority of cross-domain technology R&D investment, establish a network of manufacturing innovation research institutes, promote industrial and university cooperation in advanced manufacturing research, establish an environment to promote the commercialization of advanced manufacturing technology, and establish a national advanced manufacturing portal, etc.
Guaranteeing talent transferring	Changing public misconceptions about manufacturing, utilizing veteran talent pool, investing in community university level education, developing partnerships to provide skills certification, strengthening university projects in advanced manufacturing industries, launching key manufacturing scholarships and internship programs, etc.

Unit 1 National Strategy in the Background of Industrial Revolution

工业变革背景下的国家战略

(续)

Name	Contents
Improving business environment	Promulgate tax reform, rationalize regulatory policies, improve trade policies, renew energy policies, etc.

The report points out the specific measures of three priority areas for the development of the United States, as shown in Table 1-2.

Table 1-2　Mesures in Three Priority Areas

Technical field	Measures
Advanced sensing, control and platform systems in manufacturing industry	Establishment of manufacturing technology test bed, establishment of a research institute focusing on energy optimization and development of new industrial standards
Virtualization, information technology and digital manufacturing	Establishment of manufacturing excellence center, creation of big data manufacturing innovation research, development of CPS security and data exchange manufacturing policy standards
Advanced material manufacturing	Promoting the center for excellence in material manufacturing, managing defense assets by supply chain, developing digital standards for material design, and establishing innovation scholarships for manufacturing industry

Vocabulary　词汇

1. vital　　　　　　　['vaɪtl]　　　　　　　adj. 至关重要的；生死攸关的；有活力的
2. economy　　　　　[ɪ'kɒnəmi]　　　　　　n. 经济；节约；理财
3. continuously　　　[kən'tɪnjuəsli]　　　　adv. 连续不断地
4. replace　　　　　　[rɪ'pleɪs]　　　　　　v. 取代，代替；替换
5. sector　　　　　　　['sektə]　　　　　　　n. 部门；分区；扇区
6. revitalization　　　[ˌriːˌvaɪtəlaɪ'zeɪʃn]　　n. 恢复元气
7. pillar　　　　　　　['pɪlə(r)]　　　　　　n. 柱子，支柱
8. formulate　　　　　['fɔːmjuleɪt]　　　　　v. 制订；规划；构想；准备；认真阐述
9. institute　　　　　　['ɪnstɪtjuːt]　　　　　n. (教育、专业等) 机构，机构建筑
10. misconception　　[ˌmɪskən'sepʃn]　　　n. 错误认识；误解
11. virtualization　　　['vɜːtʃʊəˌlaɪzeɪʃn]　　n. 虚拟化
12. sense　　　　　　['sens]　　　　　　　v. 感觉到，察觉出；(机器设备) 检测出
13. digital　　　　　　['dɪdʒɪtl]　　　　　　adj. 数字的

Notes　注释

1. job-multiplier effect　　　　　　　　　　就业乘数效应

2. technology-intensive 技术密集型
3. federal government 联邦政府
4. technological innovation 技术创新
5. talent transferring 人才输送
6. business environment 商业环境
7. skills certification 技能认证
8. test bed 测试床，试验台
9. energy optimization 能源优化
10. industrial standard 工业标准
11. advanced manufacturing technology 先进制造技术
12. manufacturing industry 制造工业
13. digital manufacturing 数字制造
14. advanced material manufacturing 先进材料制造
15. big data 大数据

Reference Translation 参考译文

1. 美国制造业与竞争力

美国制造业在美国经济中发挥着至关重要的作用。制造业的进步使经济持续改善，因为新技术和创新能提高生产力，使新产品得以面世和创造全新的产业。这些新兴产业往往创造新的就业机会，从而取代低技能的工作岗位。

2. 先进制造业国家战略

2014年10月，美国发布《振兴美国先进制造业》报告，用于指导联邦政府支持先进制造研究开发的各项计划和行动。三个主要方面的具体内容见表1-1。

表1-1 三个主要方面的具体内容

名称	内容
加快技术创新	制定国家先进制造业战略，增加优先的跨领域技术的研发投资，建立制造创新研究机构网络，促进产业界和大学合作进行先进制造业方面的研究，建立促进先进制造业技术商业化的环境，建立国家先进制造业门户网站等
确保人才输送	改变公众对制造业的误解，利用资深人才库，投资社区大学水平的教育，发展伙伴关系提供技能认证，加强先进制造业的大学项目，推出关键制造业奖学金和实习计划等
改善商业环境	颁布税收改革、合理化监管政策、完善贸易政策、更新能源政策等

该报告指出了美国发展的三大优先领域的措施建议，具体内容见表1-2。

表1-2 三大优先领域具体内容

技术领域	措施
制造业中先进的传感、控制和平台系统	建立制造技术实验台，建立聚焦于能源优化利用的研究所，制定新的产业标准

Unit 1 National Strategy in the Background of Industrial Revolution
工业变革背景下的国家战略

(续)

技 术 领 域	措 施
虚拟化、信息化和数字制造	建立制造卓越能力中心，建立大数据制造创新研究，制定 CPS 安全和数据交换的制造政策标准
先进材料制造	推广材料制造卓越能力中心，利用供应链管理国防资产，制定材料设计数字标准，设立制造业创新奖学金

Part 3 Intelligent Manufacturing in China
中国智能制造

1. Background

China is facing issues such as lower growth rate of economy, cost of population aging, shrinking workforce and other social welfare system. China is also competing in the manufacturing space from newly emerging economies like Vietnam and highly industrialized countries. China needs to continuously improve its economic and technological competitiveness.

Video 3

In 2015, the "Made in China 2025" was officially promulgated by the State Council of China. The goal is to improve the competitiveness of China's manufacturing industry in the world.

2. Working Towards "Industry 4.0"

China will transform itself from a manufacturing giant into a global manufacturing power in the next three decades. "Made in China 2025" is the first 10-year plan of the Three-Step Strategy, which aims to uplift China to the second group of global manufacturing bases from the third one, and ultimately into the first group by 2045. The content of "Made in China 2025" is shown in Figure 1-2.

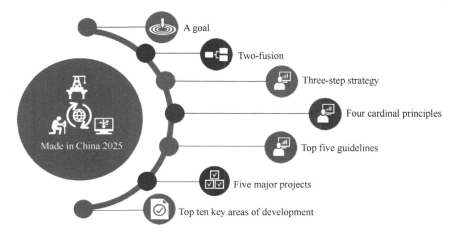

Figure 1-2 The Content of "Made in China 2025"

3. Five Major Projects

"Made in China 2025" will focus on five major projects, as shown in Figure 1-3.

(1) Manufacturing innovation center Focus on basic and common key technology research and development, industrialization of achievements, talent training and other work.

(2) Intelligent manufacturing Develop the integration innovation and engineering application of new generation information technology and manufacturing equipment, establish the standard system of intelligent manufacturing and information security system, etc.

(3) Industrial strong foundation Focus on key basic materials, core basic components, advanced basic technology and industrial technology.

(4) Green manufacturing Organize and implement the special technical transformation of energy efficiency improvement, cleaner production, water saving and pollution control in traditional manufacturing industry, and formulate the standard system of green products, green factories and green enterprises.

(5) High-end equipment Organize and implement a number of innovative and industrialized projects such as large aircraft, aeroengine, smart grid and high-end medical equipment.

Figure 1-3 Five Major Projects of "Made in China 2025"

Vocabulary　词汇

1. emerging	[ɪˈmɜːdʒɪŋ]	adj.	新兴的；出现的；形成的
2. aerospace	[ˈeərəʊspeɪs]	n.	航空宇宙；航空航天空间
3. competitiveness	[kəmˈpetətɪvnɪs]	n.	竞争力
4. degradation	[degrəˈdeɪʃən]	n.	退化；降格，降级
5. promulgate	[ˈprɒməlgeɪt]	v.	公布；传播；发表
6. semiconductor	[semɪkənˈdʌktə]	n.	半导体
7. ultimately	[ˈʌltɪmətlɪ]	adv.	最后；根本；基本上
8. robotics	[rəʊˈbɒtɪks]	n.	机器人科学（或技术）

Unit 1　National Strategy in the Background of Industrial Revolution
工业变革背景下的国家战略

| 9. equipment | [ɪˈkwɪpmənt] | n. 设备；器材；配备；装备 |
| 10. automotive | [ˌɔːtəˈməʊtɪv] | adj. 汽车的；机动车辆的 |

Notes　注释

1. population aging　　　　　　　人口老龄化
2. high-tech fields　　　　　　　高新技术领域
3. standard of living　　　　　　生活标准
4. Made in China 2025　　　　　中国制造 2025
5. intelligent manufacturing　　　智能制造
6. manufacturing base　　　　　　制造基地
7. information security system　　信息安全系统
8. green manufacturing　　　　　绿色制造
9. high-end equipment　　　　　　高端装备

Reference Translation　参考译文

1. 背景

中国正面临着经济增长放缓、人口老龄化、劳动力萎缩等诸多内部问题。在制造领域，中国还要与新兴的经济体比如越南和其他高度工业化的国家竞争，中国需要不断提升经济和技术竞争力。

国务院在 2015 年正式发布《中国制造 2025》，目标是提高中国制造业在世界上的竞争力。

2. 迈向"工业 4.0"

未来 30 年，中国将从制造业大国转变为全球制造业强国。《中国制造 2025》是三步走战略的第一个 10 年计划，其目标是将中国从全球制造业基地的第三方阵提升到第二方阵，并最终在 2045 年进入第一方阵。《中国制造 2025》的内容如图 1-2 所示。

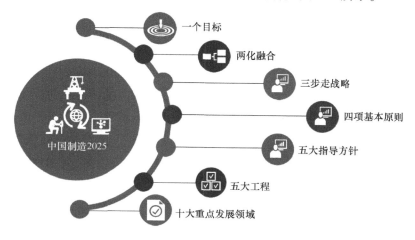

图 1-2　《中国制造 2025》的内容

3. 五大工程

《中国制造 2025》将重点实施五大工程，如图 1-3 所示。

（1）制造业创新中心建设工程　重点开展行业基础和共性关键技术研发、成果产业化、人才培训等工作。

（2）智能制造工程　开展新一代信息技术与制造装备融合的集成创新和工程应用；建立智能制造标准体系和信息安全保障体系等。

（3）工业强基工程　以关键基础材料、核心基础零部件（元器件）、先进基础工艺、产业技术基础为发展重点。

（4）绿色制造工程　组织实施传统制造业能效提升、清洁生产、节水治污等专项技术改造；制定绿色产品、绿色工厂、绿色企业标准体系。

（5）高端装备创新工程　组织实施大型飞机、航空发动机、智能电网、高端医疗设备等一批创新和产业化专项、重大工程。

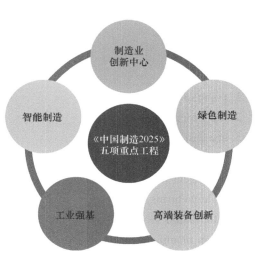

图1-3　《中国制造2025》五项重点工程

Part 4　Other National Strategies 其他国家战略

1. New Industrial Program in France

The New Industrial Program in France was launched on April 18, 2015. Its goal is to modernize France's production tools and provide support for manufacturers as the digital changeover transforms their business models, organizations and the way they design and market their products. The New Industrial France Program is based on nine industrial solutions, as shown in Figure 1-4.

Video 4

2. British National Manufacturing Policy

In October 2013, the UK launched the UK Industrial 2050 Strategy, formulated the future manufacturing development strategy by 2050, and put forward the UK manufacturing development and recovery policy. The report identifies four characteristics of future British manufacturing, as shown in Figure 1-5.

1) Respond quickly and acutely to consumer demand.
2) Seize new market opportunities.
3) The sustainable development of manufacturing industry.
4) Skilled workers.

3. National Manufacturing Policy in India

Industrial development in India was catalyzed by three major industrial policy resolutions of Indian Government in 1948, 1956 and 1991. Economic reforms unveiled in 1991 have brought about a

Unit 1 National Strategy in the Background of Industrial Revolution
工业变革背景下的国家战略

Figure 1-4 New Industrial Program in France

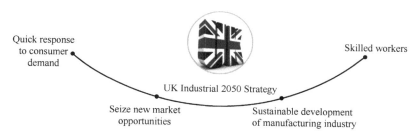

Figure 1-5 Four Characteristics of Future British Manufacturing

structural shift enabling the private sector to assume a much larger role in all sectors of economy. In the last two decades, Indian economy has witnessed a transformational change and has emerged as one of the fastest growing economies of the world.

4. Japan's New Robot Strategy

Japan's excellence in the field of manufacturing has particularly been notable in the area of industrial robots. Japan has maintained its global position as the world's number one supplier of industrial robots. In addition, Japan holds great share of over 90% worldwide in the field of key parts for robots such as precision reduction gear for robots, servo motor and force sensor.

Vocabulary 词汇

1. modernize	[ˈmɒdənaɪz]	v.	使……现代化
2. changeover	[ˈtʃeɪndʒəʊvə(r)]	n.	转换；逆转
3. catalyze	[ˈkætəlaɪz]	v.	催化；刺激，促进
4. unit	[ˈjuːnɪt]	n.	单位
5. precision	[prɪˈsɪʒn]	n.	精度，精密度；精确
6. reduction	[rɪˈdʌkʃn]	n.	减少；下降；缩小；还原反应

7. excellence	[ˈɛksləns]	n. 优秀；美德；长处
8. witness	[ˈwɪtnəs]	v. 见证
9. part	[pɑːt]	n. 部分；零件

Notes 注释

1. private sector　　　　私营部门
2. in the area of　　　　在……领域
3. reduction gear　　　　减速齿轮
4. servo motor　　　　　伺服电动机
4. force sensor　　　　　力传感器
6. smart object　　　　　智能对象
7. new resource　　　　　新能源
8. industrial robot　　　　工业机器人

Reference Translation 参考译文

1. 法国新工业计划

法国新工业计划于 2015 年 4 月 18 日启动。其目标是使法国的生产工具现代化，并在数字浪潮改变企业的商业模式、组织及其设计和营销产品的方式时，为企业提供支持。法国新工业计划以 9 项工业解决方案为基础，如图 1-4 所示。

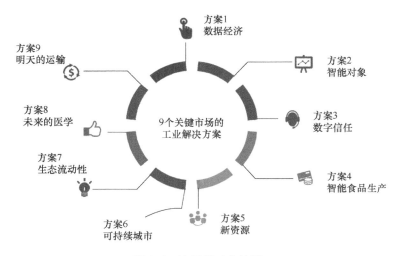

图 1-4　法国新工业计划

2. 英国国家制造政策

2013 年 10 月英国推出了《英国工业 2050 战略》，制定了到 2050 年的未来制造业发展战略，提出了英国制造业发展与复苏政策。报告确定了未来英国制造业的四个特点，如图 1-5 所示。

1）快速、敏锐地响应消费者需求。
2）把握新的市场机遇。

3）可持续发展的制造业。
4）培养高技能工作者。

图1-5 未来英国制造业的特点

3. 印度国家制造政策

印度的工业发展是由印度政府在1948年、1956年和1991年的三次工业政策改革推动的。印度1991年的经济改革带来了结构性转变，使私营部门能够在所有经济部门中发挥更大的作用。在过去的20年里，印度经济经历了一场转型并已成为世界上增长最快的经济体之一。

4. 日本新机器人战略

日本在制造业领域的卓越表现是工业机器人领域。日本一直保持着其作为世界第一的工业机器人供应商的地位。此外，日本在精密减速器、伺服电动机和力传感器等机器人关键零部件领域的全球份额超过90%。

Unit 2 Theme of Intelligent Manufacturing

智能制造的主题

Part 1　Intelligent Factory
智能工厂

1. Introduction

Intelligent factory uses various modern technologies to realize the automation of office, management and production, as Shown in Figure 2-1. The goal of building intelligent factory is to strengthen and standardize enterprise management, reduce work errors, improve work efficiency, carry out safety production and provide decision-making reference.

Video 5

2. Features

Intelligent factory is the new direction for intelligent manufacturing development. The features of intelligent factory are following:

(1) Equipment interconnection　Equipment in intelligent factories can be interconnected. Data acquisition and monitoring system collects real-time equipment status, production information and quality information, and achieves traceability of production process.

(2) Wide use of industrial software　In intelligent factories, MES(Manufacturing Execution System), APS(Advanced Production Scheduling), energy management, quality management and other industrial software are widely used.

(3) Flexible automation　Logistics automation is very important for the realization of intelligent factory. Enterprises can transfer materials between processes through AGV(Automated Guided Vehicle), truss manipulator and other logistics equipment, as shown in Figure 2-2.

(4) Green manufacturing　Intelligent factory collects energy consumption of equipment and

Unit 2 Theme of Intelligent Manufacturing
智能制造的主题

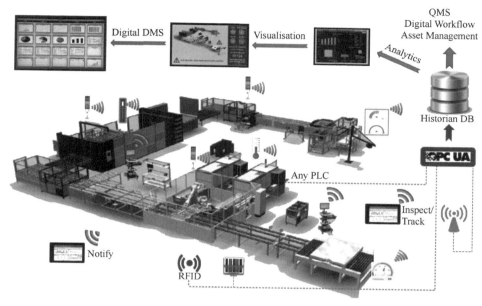

Figure 2-1　Example of Intelligent Factory

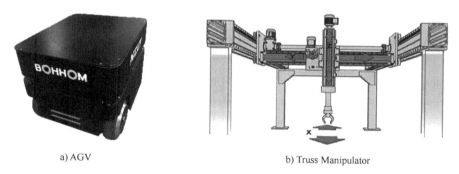

Figure 2-2　Flexible Automation

production line in time, and realizes efficient energy utilization. In dangerous and polluted links, robots are used instead of humans.

3. Architecture of Intelligent Factory

The architecture of intelligent factory can be divided into three levels, as shown in Figure 2-3.

(1) Field level　It is the lowest level of the automation hierarchy and consists of field devices such as sensors and actuators. Sensors collect data on temperature, pressure, speeds, feeds, and so on, convert it to electrical signals, and relay it up to the next level. The main task of these field devices is to transfer data on processes and machines for monitoring and analysis.

It also includes the actuators, which are controlled by the next level through electrical or pneumatic signals, converting them into actions. Actuators turn valves, relays, motors, pumps, and other devices on or off, or adjust their outputs to control the processes.

(2) Control level　This level consists of various automation controllers such as CNC(Computer

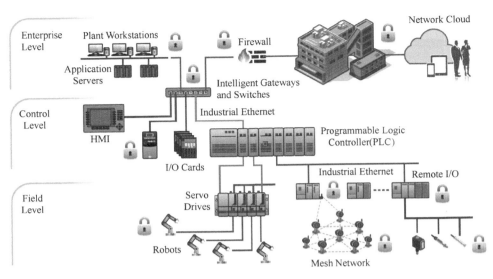

Figure 2-3   Architecture of Intelligent Factory

Numerical Control) machines that gather process parameters from various sensors. The automatic controllers then drive the actuators based on the processed sensor signals and the program or control technique.

Programmable logic controller(PLC) is the most widely used and durable industrial controller that can deliver automatic control functions based on sensor inputs. PLC consists of various modules such as the CPU, analog and digital I/Os, and communication modules.

(3) Enterprise level   This top level of industrial automation, which is also called Information-level automation, manages the whole automation system. The level handles production planning, customer and market analysis, and orders and sales. It deals more with commercial activities and less with technical aspects.

Industrial communication networks tie all these levels together, sending data from one level to the other. These communication networks can be different form level to level. With this hierarchy in place, there is a continuous flow of information from high level to low level and vice versa.

Vocabulary   词汇

1. strengthen   ['streŋθn]   v. 加强；增强；巩固
2. enterprise   ['entəpraɪz]   n. 公司；企业
3. traceability   [treɪsə'bɪlɪti]   n. 可追溯
4. integrate   ['ɪntɪgreɪt]   v. （使）合并，成为一体
5. controller   [kən'trəʊlə(r)]   n. （机器的）控制器，调节器
6. monitoring   ['mɒnɪtərɪŋ]   v. 监控
7. status   ['steɪtəs]   n. 状态，情况
8. sensor   ['sensə(r)]   n. （探测光、热、压力等的）传感器
9. actuator   ['æktjʊeɪtə]   n. 执行机构（元件）

Unit 2　Theme of Intelligent Manufacturing
智能制造的主题

Notes　注释

1. flexible automation　　　　　　柔性自动化
2. data acquisition　　　　　　　　数据采集
3. green manufacturing　　　　　　绿色制造
4. truss manipulator　　　　　　　桁架机械手
5. energy consumption　　　　　　能源消耗
6. manufacturing execution system　制造执行系统
7. advanced production scheduling　高级生产排程
8. intelligent factory　　　　　　　智能工厂
9. equipment interconnection　　　设备互联
10. industrial software　　　　　　工业软件
11. logistics automation　　　　　　物流自动化
12. field level　　　　　　　　　　现场层
13. automation hierarchy　　　　　自动化体系
14. field device　　　　　　　　　现场设备
15. control level　　　　　　　　　控制层
16. automation controller　　　　　自动化控制器
17. CNC machine　　　　　　　　　数控机床
18. programmable logic controller　可编程序控制器
19. industrial controller　　　　　　工业控制器
20. enterprise level　　　　　　　　企业级
21. production planning　　　　　　生产计划

Reference Translation　参考译文

1. 介绍

智能工厂，就是利用各种现代化的技术实现办公、管理及生产自动化，如图2-1所示。建设智能工厂的目标是加强及规范企业管理，减少工作失误，提高工作效率，开展安全生产，提供决策参考。

2. 特征

智能工厂是智能制造发展的新方向。智能工厂的特征如下：

（1）设备互联　智能工厂中的设备能够相互连接。由数据采集与监控系统实时采集设备的状态、生产信息和质量信息，实现生产过程的可追溯性。

（2）广泛应用工业软件　智能工厂广泛应用MES（制造执行系统）、APS（先进生产排程）、能源管理和质量管理等工业软件。

（3）柔性自动化　物流自动化对于实现智能工厂至关重要，企业可以通过AGV（无人搬运车）、桁架机械手等物流设备实现工序之间的物料传递，如图2-2所示。

（4）实现绿色制造　智能工厂能够及时采集设备和生产线的能源消耗，实现能源的高效利用。在危险和存在污染的环节，优先选用机器人替代人工操作。

3. 智能工厂的架构

智能工厂的架构可分为三个层次，如图 2-3 所示。

（1）现场层　它是自动化体系的最低层次，由传感器和执行器等现场设备组成。传感器收集温度、压力、速度、进料等数据，将其转换为电信号，并将其传输到控制层级。这些现场设备的主要任务是传输过程和机器上的数据从而进行监视和分析。

现场层还包括执行器，由控制层级通过电气或气动信号控制，将信号转换为动作。执行器打开或关闭阀门、继电器、电动机、泵和其他设备，或调整其输出以控制过程。

（2）控制层　控制层由各种自动化控制器组成，例如从各种传感器收集工艺参数的数控机床（CNC）。然后，自动控制器根据处理后的传感器信号和程序或控制技术驱动执行器。

可编程序控制器（PLC）是应用最广泛、最耐用的工业控制器，能够根据传感器输入提供自动控制功能。它们由各种模块组成，如 CPU、模拟和数字 I/O 以及通信模块。

（3）企业层　企业层位于工业自动化的顶层，也称为信息级自动化，管理着整个自动化系统。该级别负责生产计划、客户和市场分析、订单及销售。它更多地涉及商业活动，而较少涉及技术方面。

工业通信网络将所有这些层级连接在一起，将数据从一个层级发送到另一个层级。这些通信网络可以是不同形式的层级。有了这个三层的层级结构，就有了从高层级到低层级的连续信息流，以及从低层级到高层级的连续信息流。

Part 2　Intelligent Production
智能生产

1. Definition

Intelligent production is a human-machine integration system composed of intelligent equipment, sensors, process control, intelligent logistics, manufacturing execution system, and cyber-physical system. The goal of intelligent production is to optimize the productivity of manufacturing systems while considering maintenance of the equipment and energy usage. Figure 2-4 showns an example architecture of intelligent production. The architecture includes different modular models. A large amount of operational data that is being produced by the manufacturing process is pushed to the cloud, and the manufacturing process is analyzed automatically in real time. Data from both cyber and physical domains are used to develop an anomaly detection framework. Optimization functions can encode the tradeoffs between productivity, energy usage, and reliability, and solve for the optimal machine parameters.

Video 6

2. Composition

The compositions of the intelligent production system include:

（1）Digital and intelligent equipment　In order to meet the needs of personalized customiza-

Unit 2　Theme of Intelligent Manufacturing
智能制造的主题

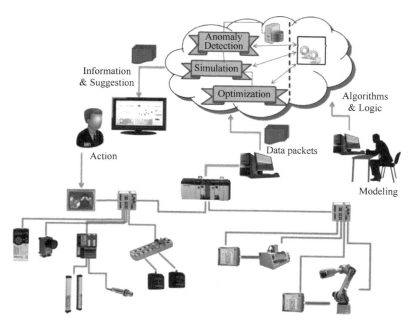

Figure 2-4　Example Architecture of Intelligent Production

tion, manufacturing equipment should be digitized and intelligent. Intelligent production system should include a few of flexible manufacturing systems. Each flexible manufacturing system independently completes manufacturing processes such as processing, assembly and welding of components.

（2）Intelligent warehouse and logistics　Intelligent warehouse should be constructed to handle materials. Material stacking and distribution can be automatically completed by the intelligent logistics system, as shown in Figure 2-5.

Figure 2-5　Intelligent Warehouse and Logistics

（3）Intelligent Manufacturing execution management(MES)　The functions of the MES include planning and distribution of workshop jobs, process monitoring, material tracking and management, operation and monitoring of workshop equipment, etc.

Vocabulary 词汇

1. composition [ˌkɒmpəˈzɪʃn] n. 成分；构成；组合方式
2. independently [ˌɪndɪˈpendəntlɪ] adv. 独立地；自立地
3. construct [kənˈstrʌkt] v. 组成；创建
4. assembly [əˈsemblɪ] n. 装配
5. welding [ˈweldɪŋ] v. 焊接

Notes 注释

1. human-machine integration 人机一体化
2. intelligent logistic 智能物流
3. intelligent warehouse 智能仓库
4. process monitoring 过程监控
5. material tracking 物料跟踪
6. intelligent production 智能生产
7. intelligent equipment 智能装备
8. process control 过程控制
9. production line 生产线
10. industrial Ethernet 工业以太网
11. industrial wireless network 工业无线网络

Reference Translation 参考译文

1. 定义

智能生产就是使用智能装备、传感器、过程控制、智能物流、制造执行系统、网络物理系统组成的人机一体化系统。智能生产的目标是优化制造系统的生产率，同时考虑设备的维护和能源使用。图 2-4 显示了智能生产的一个示例架构。该体系结构包括不同的模块化模型。通过将制造过程中产生的大量操作数据推送到云端，可以对制造过程进行实时的自动分析。使用来自网络和物理域的数据来开发异常检测框架。优化函数可以对生产率、能量利用率和可靠性进行平衡，从而求解最优的机器参数。

2. 组成

智能生产系统的组成包括以下三个方面：

（1）数字智能设备　为了满足个性化定制的需要，制造设备必须数字化、智能化。智能生产系统应包括一些柔性制造系统。各柔性制造系统独立完成零部件的加工、装配、焊接等制造过程。

（2）智能仓库与物流　应建造自动化仓库来处理物料，由智能物流系统来自动完成物料的堆放和配送，如图 2-5 所示。

（3）智能制造执行系统（MES）　智能制造执行系统的功能包括车间作业的规划和分配、过程监控、物料跟踪和管理、车间设备的运行和监控等。

Unit 2 Theme of Intelligent Manufacturing
智能制造的主题

Intelligent Logistics
智慧物流

1. Definition

Intelligent logistics applies advanced Internet of Things technology to the basic process of logistics industry, such as transportation, warehousing, distribution, packaging, loading and unloading, as shown in Figure 2-6. The goal of intelligent logistics is to realize the automation operation and optimal management of cargo transportation process, improve the service level of logistics industry, reduce costs, and reduce the consumption of natural resources and social resources.

Video 7

2. Features

The characteristics of intelligent logistics are shown in Figure 2-7.

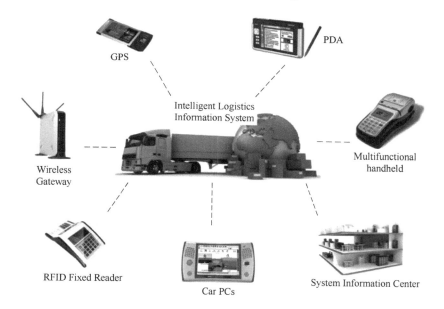

Figure 2-6　Example Components of Intelligent Logistics Information System

(1) Intelligent　Intelligent logistics system, using database and data analysis technology, can respond to logistics changes and take corresponding actions according to different situations.

(2) Integration and hierarchy　Intelligent logistics system takes logistics management as the center, realizes the integration of transportation, storage, packaging, loading and unloading, and the hierarchy of intelligent logistics system in the logistics process.

(3) Flexibility　With the development of e-commerce, the producer-centered business model has changed to consumer-centered business model. Intelligent logistics needs to adjust the process

Figure 2-7　Features of Intelligent logistics

according to the needs of consumers, so as to realize the flexibility of logistics system.

(4) Socialization　The development of intelligent logistics will drive the development of regional economy and Internet economy, changing people's life style in some aspects.

Vocabulary　词汇

1. logistics	[lə'dʒɪstɪks]	n. 物流；组织工作
2. optimal	['ɒptɪməl]	adj. 最优的，最佳的
3. cargo	['kɑːgəʊ]	n. （船或飞机装载的）货物
4. emphasize	['emfəsaɪz]	v. 强调；重视；着重
5. integration	[ˌɪntɪ'greɪʃn]	n. 结合；整合；一体化
6. transportation	[ˌtrænspɔː'teɪʃn]	n. 运输
7. warehousing	['weəhaʊzɪŋ]	n. 仓储
8. packaging	['pækɪdʒɪŋ]	n. 包装
9. handheld	['hændheld]	n. 手持式的，便携的

Notes　注释

1. logistic process　　　物流过程
2. regional economy　　 区域经济
3. logistics industry　　 物流工业
4. cargo transportation　货物运输
5. wireless gateway　　 无线网关
6. data analysis　　　　数据分析

Reference Translation　参考译文

1. 定义

智慧物流将先进的物联网技术应用于物流业基本环节，比如运输、仓储、配送、包装和装卸等，如图 2-6 所示。智慧物流的目标是实现货物运输过程的自动化运作和优化管理，提高物流行业的服务水平，降低成本，减少自然资源和社会资源消耗。

2. 智慧物流的特点

智慧物流的特点如下：

（1）智能化　智慧物流系统运用数据库和数据分析技术，对物流变化具有响应机制，可以针对不同的情况采取相应的措施。

（2）一体化和层次化　智慧物流系统以物流管理为中心，实现物流过程中运输、存储、包装和装卸等环节的一体化和智慧物流系统的层次化。

（3）柔性化　随着电子商务的发展，以生产商为中心的商业模式转变为以消费者为中心的商业模式。智慧物流需要根据消费者的需求来调整流程，从而实现物流系统的柔性化。

（4）社会化　智慧物流会带动区域经济和互联网经济的高速发展，从而在某些方面改变人们的生活方式。

Unit 3 Key Technology of Intelligent Manufacturing

智能制造关键技术

Part 1 Artificial Intelligence
人工智能

Video 8

Artificial intelligence(AI), known as machine intelligence, refers to the ability of computers to perform human-like feats of cognition including learning, problem-solving, perception, decision-making, and speech and language. AI is widely used in self-driving cars, improved fraud detection, and "personal assistants" like Siri and Alexa. As shown in Figure 3-1, artificial intelligence generally includes six key technologies.

At present, artificial intelligence technology has greatly changed various industries. For example, in agriculture, industry and other fields, specialized knowledge can now be grasped by ordinary people and guide their production practice. AI technology also plays an assistant role in medical, aerospace and other highly specialized fields. Online Car-hailing, targeting, elderly care are new applications of artificial intelligence technology. The main application areas of AI are shown in Figure 3-2.

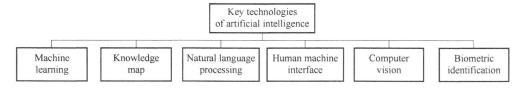

Figure 3-1 Key Technologies of Artificial Intelligence

Unit 3 Key Technology of Intelligent Manufacturing

智能制造关键技术

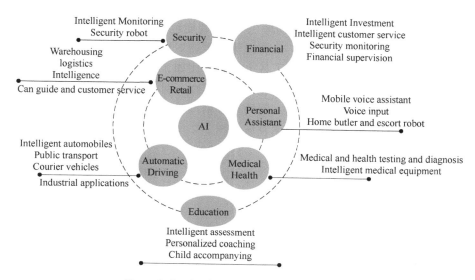

Figure 3-2 Artificial Intelligence Applications

Vocabulary 词汇

1. cognition　　　　　[kɒgˈnɪʃn]　　　　　n. 认知；感知；认识
2. perception　　　　[pəˈsepʃn]　　　　　n. 感知；洞察力；看法；见解
3. algorithm　　　　　[ˈælgərɪðəm]　　　　n. 算法；计算程序
4. storage　　　　　　[ˈstɔːrɪdʒ]　　　　　n. 贮存，贮藏；存储
5. sophisticated　　　[səˈfɪstɪkeɪtɪd]　　　adj. 复杂巧妙的；先进的；精密的
6. grasp　　　　　　　[grɑːsp]　　　　　　v. 理解；领会；领悟；明白

Notes 注释

1. fraud detection　　　　　　　欺诈检测
2. biometric identification　　　生物识别
3. artificial intelligence　　　　人工智能
4. machine intelligence　　　　机器智能
5. decision-making　　　　　　决策
6. data volume　　　　　　　　数据容量
7. self-driving car　　　　　　　自动驾驶汽车
8. knowledge map　　　　　　　知识图谱
9. natural language processing　自然语言处理
10. human machine interface　　人机接口
11. computer vision　　　　　　机器视觉

Reference Translation 参考译文

人工智能（AI）被称为机器智能，是指计算机能够执行类似人类的认知活动的能力，

包括学习、解决问题、感知、决策以及言语和语言。其主要应用有自动驾驶车辆、改进的欺诈检测以及 Siri 和 Alexa 这样的"个人助理"。人工智能包含了6项关键技术，即机器学习、知识图谱、自然语言处理、人机接口、机器视觉和生物识别，如图3-1所示。

目前，人工智能技术已极大地改变了各个行业。比如，在农业、工业等领域，原本专业化的知识可以为普通人所掌握，指导他们进行生产实践。对于医疗、航天等专业性极强的领域，人工智能技术也可以起到辅助作用。网约车、精准推送、老年人护理等领域，更是人工智能技术的新应用。人工智能的主要应用领域如图3-2所示。

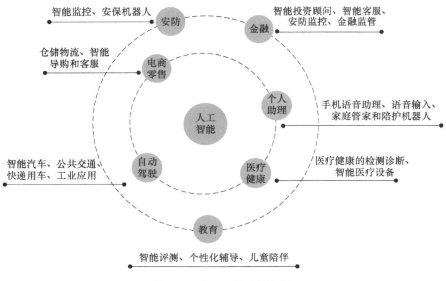

图 3-2　人工智能应用领域

Part 2　Industrial Internet of Things
工业物联网

1. Definition

The Industrial Internet of Things(IIOT) refers to interconnected sensors, instruments, and other devices networked together with computers' industrial applications. This connectivity allows for data collection, exchange and analysis, potentially facilitating improvements in productivity and efficiency.

Video 9

The IIoT is an evolution of a Distributed Control System(DCS) that allows for a higher degree of automation by using cloud computing to refine and optimize the process controls.

2. Enabling Technologies

The IIoT is enabled by many technologies, three of the most important ones are described as below.

Unit 3 Key Technology of Intelligent Manufacturing

(1) Cloud computing With cloud computing, IT services can be delivered in which resources are retrieved from the Internet as opposed to direct connection to a server. Files can be kept on cloud-based storage systems rather than on local storage devices. As shown in Figure 3-3.

(2) Big data analytics Big data analytics is the process of examining large and varied data sets, or big data. As shown in Figure 3-4.

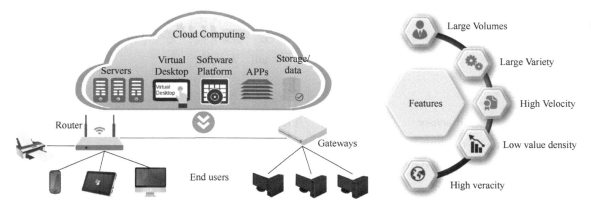

Figure 3-3 Cloud Computing Conceptual Model　　　　Figure 3-4 Big Data 5V Features

(3) Wireless sensor network Wireless Sensor Network(WSN) refers to a group of spatially dispersed and dedicated sensors for monitoring and recording the physical conditions of the environment and organizing the collected data at a central location. WSNs measure environmental conditions like temperature, sound and so on. Composition of wireless sensor network system is shown in Figure 3-5. Today WSNs are used in many industrial and consumer applications, such as industrial process monitoring and control and so on.

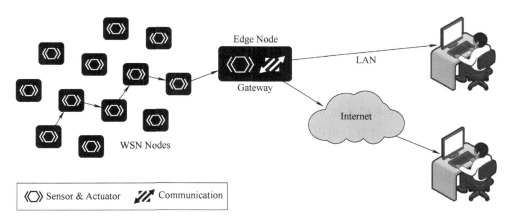

Figure 3-5 Composition of Wireless Sensor Network System

In addition, IIOT involves mobile technologies, machine-to-machine, 3D printing, advanced robotics, cyber-physical systems, RFID technology, and cognitive computing.

Vocabulary 词汇

1. connectivity　　[ˌkɒnek'tɪvəti]　　n. 连通性
2. evolution　　[ˌiːvə'luːʃn]　　n. 进化；演变；发展；渐进
3. motivate　　['məʊtɪveɪt]　　v. 成为……的动机；激励；激发
4. instrument　　['ɪnstrəmənt]　　n. 器械；仪器
5. optimize　　['ɒptɪmaɪz]　　v. 使最优化；充分利用
6. wireless　　['waɪələs]　　n. 无线电报；无线电发射和接收系统
　　　　　　　　　　　　　　 adj. 无线的

Notes 注释

1. interconnected sensor　　互联传感器
2. machine-to-machine　　机器对机器
3. distributed control system　　分布式控制系统
4. data collection　　数据采集
5. mobile technology　　移动技术
6. cognitive computing　　认知计算
7. wireless sensor network　　无线传感网络
8. cloud-based storage system　　基于云的存储系统
9. big data analytics　　大数据分析

Reference Translation 参考译文

1. 定义

工业物联网（IIOT）是指与计算机的工业应用联网的传感器、仪器和其他设备。通过这种相互连接，可以进行数据采集、交流和分析，从而促进生产力和效率的提高。IIoT 是分布式控制系统（DCS）的一个发展，它允许通过使用云计算技术来细化和优化过程控制。

2. 使能技术

IIoT 三个最重要的技术描述如下：

（1）云计算　通过云计算，可以通过从 Internet 检索资源提供 IT 服务，而不需要直接连接到服务器。文件可以保存在基于云的存储系统上，而不是本地存储设备上。

（2）大数据分析技术　大数据分析技术是分析各种大规模数据集的过程。

（3）无线传感网络　无线传感网络是指一组在空间上分散且专用的传感器，用于监视和记录环境的物理条件并在中心位置组织所收集的数据。无线传感网络测量环境条件，如温度、声音、污染程度、湿度和风力等。目前，这种网络被用于许多工业和消费应用中，例如工业过程监视和控制等。

此外，IIOT 还涉及移动技术、机器对机器、3D 打印、先进机器人学、网络物理系统、射频识别技术和认知计算等技术。

Unit 3 Key Technology of Intelligent Manufacturing

智能制造关键技术

Part 3 Internet of Things and Cyber-Physical System
物联网与网络物理系统

1. What is the Internet of things?

According to the definition of the International Telecommunication Union(ITU), the Internet of Things(IOT) is a ubiquitous network of intelligent identification, positioning, tracking, monitoring and management through information sensing devices that connect any item to the Internet, in accordance with an agreement, for information exchange and communication. Several features of the IoT are shown in figure 3-6.

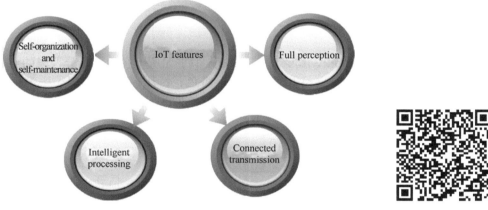

Figure 3-6 Features of Internet of Things

Video 10

2. Cyber-Physical Systems

The Cyber-Physical System(CPS) is a multidimensional and complex system which integrates the computing, communication and control technology. CPS realizes the real-time perception of large engineering systems through the fusion and deep collaboration of 3C - Computing, Communication and Control technology, as shown in Figure 3-7.

The information world refers to industrial software and management software, industrial design, the internet and mobile internet, etc. The physical world refers to the energy environment, human, work environment, local communications, as well as equipment and products.

Future CPS need to be scalable, distributed, decentralized allowing in-

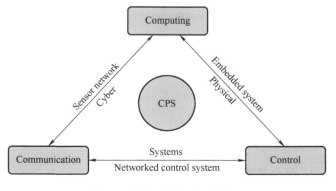

Figure 3-7 CPS's Core"3C"

teraction with humans, environment and machines. Adaptability, reactivity, optimality and security are features to be embedded in such systems, as shown in Figure 3-8 and Figure 3-9.

Figure 3-8 Example of Smart Home

Figure 3-9 Example of Intelligent Transportation

Vocabulary　词汇

1. ubiquitous　　　　[juːˈbɪkwɪtəs]　　　　adj. 似乎无所不在的；十分普遍的
2. feature　　　　　[ˈfiːtʃə(r)]　　　　　n. 特色；特征；特点 v. 以……为特色
3. multidimensional　[ˌmʌltɪdɪˈmenʃənl]　　adj. 多维的；多面的
4. fusion　　　　　 [ˈfjuːʒn]　　　　　　n. 融合；结合
5. identification　　[aɪˌdentɪfɪˈkeɪʃn]　　　n. 识别；确定
6. tracking　　　　 [ˈtrækɪŋ]　　　　　　v. 跟踪；追踪
7. computing　　　 [kəmˈpjuːtɪŋ]　　　　 n. 计算 v. 计算；估算
8. communication　 [kəˌmjuːnɪˈkeɪʃn]　　　n. 交流；传递；通信
9. scalable　　　　 [ˈskeɪləbl]　　　　　　adj. 可扩展的
10. distributed　　　[dɪˈstrɪbjuːtɪd]　　　　adj. 分布的；分散的 v. 分发；分配
11. decentralized　　[ˌdiːˈsentrəlaɪzd]　　　v. 分散；去中心化
12. optimality　　　[ɔptiˈmæliti]　　　　　n. 最优性

Notes　注释

1. according to　　　　　　　　　　　　根据
2. International Telecommunication Union　国际电信联盟
3. intelligent identification　　　　　　　 智能识别
4. information sensing device　　　　　　信息传感设备
5. information exchange and communication　信息交换与通信
6. physical environment　　　　　　　　物理环境
7. real-time perception　　　　　　　　 实时感知

Reference Translation　参考译文

1. 什么是物联网？

根据国际电信联盟（ITU）的定义，物联网（IOT）是一个通过信息传感设备，按照约

Unit 3 Key Technology of Intelligent Manufacturing

定的协议，把任何物品与互联网相连接，进行信息交换和通信，以实现智能化识别、定位、跟踪、监控和管理的泛在网络。物联网的几个特点如图3-6所示。

2. 网络物理系统

网络物理系统（CPS）是一个集计算、通信和控制技术于一体的多维复杂系统。该系统通过3C（计算、通信和控制技术）的融合与深度协作，可实现大型工程系统的实时感知，如图3-7所示。

信息世界是指工业软件和管理软件、工业设计、互联网和移动互联网等；物理世界是指能源环境、人、工作环境、局域通信以及设备与产品等。

未来的网络物理系统需要具有可扩展性、分布式、分散性，以便与人类、环境和计算机进行交互。适应性、反应性、优化性和安全性等特征也需要被嵌入此类系统中。

图 3-6 物联网的特点

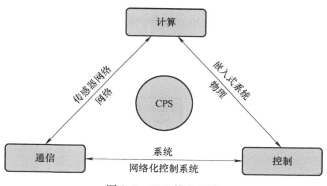

图 3-7 CPS 核心元素

Part 4 Other Technologies
其他技术

1. Virtual Reality(VR)

Virtual reality(VR) is an interactive computer-generated experience taking place within a simulated environment. It incorporates mainly auditory and visual feedback, but may also allow other types of sensory feedback.

There are three characteristics of virtual reality, as shown in Figure 3-10.

(1) Immersion refers to the use of three-dimensional images to give people a feeling of being present on the scene.

(2) Interactivity refers to the use of sensor devices to interact with user, so that users feel as if they are in the real world.

(3) Imagination refers to enabling users to immerse themselves in the virtual environment and improve their perceptual and rational under-

Video 11

standing, thus enabling users to derive new ideas and concepts in cognition.

The general virtual reality system mainly includes five components: professional graphics processing computer, application software system, input device, output device and database, as shown in Figure 3-11.

2. 3D printing

3D printing is a process in which material is joined or solidified under computer control to create a three-dimensional object. The principle of 3D printing technology is shown in Figure 3-12. Firstly, a series of thin slices (usually 0.01 ~ 0.1mm) are obtained by intercepting the part's digital three-

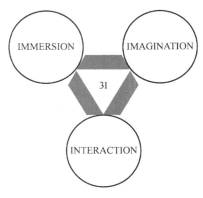

Figure 3-10 Characteristics of Virtual Reality

dimensional CAD model by a set of parallel planes, and then the whole part can be accumulated by the thin slices according to certain rules.

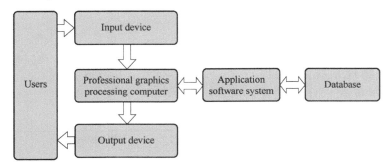

Figure 3-11 Example Composition of A Virtual Reality System

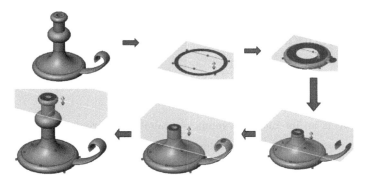

Figure 3-12 Schematic Diagram of the Principle of 3D Printing Technology

Today, with the increase in precision, repeatability and material range, 3D printing is considered as an industrial production technology, with the name of additive manufacturing. 3D printed objects can have a very complex shape or geometry.

3. Radio Frequency Identification(RFID)

Radio Frequency Identification(RFID) is one method of automatic identification and data cap-

Unit 3 Key Technology of Intelligent Manufacturing

智能制造关键技术

ture. RFID uses electromagnetic fields to automatically identify and track tags attached to objects, which contain electronically stored information.

The composition of RFID system varies from application to application, but it is basically composed of electronic tag, reader and data management system, as shown in Figure 3-13.

(1) Electronic tag It has the functions of intelligent reading and writing and encrypted communication. It exchanges data with reading and writing equipment through radio waves. The energy is provided by radio frequency pulses emitted by the reader.

(2) Reader The reader can transmit the read and write commands of the host computer to the electronic label through the antenna, encrypt the data sent from the host computer to the electronic label, decrypt the data returned from the electronic label and send it to the host computer.

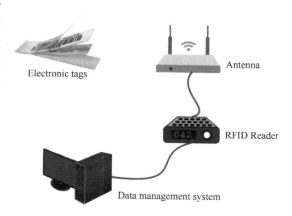

Figure 3-13 Composition of RFID System

(3) The data management system It is responsible for the storage and management of data information, and the control of reading/writing of cards, etc.

Vocabulary　词汇

1. simulate	['sɪmjuleɪt]	v.	模拟；模仿
2. sensory	['sensəri]	adj.	感觉的；感官的
3. demonstrate	['demənstreɪt]	v.	证明；证实；论证；说明
4. extinguish	[ɪk'stɪŋgwɪʃ]	v.	熄灭；扑灭
5. automobile	['ɔːtəməbiːl]	n.	汽车
6. merchandise	['mɜːtʃəndaɪs]	n.	商品；货品
7. solidified	[sə'lɪdɪfaɪd]	v.	（使）凝固，变硬
8. geometry	[dʒi'ɒmətri]	n.	几何（学）；几何形状
9. interrogating	[ɪn'terəgeɪtɪŋ]	v.	（在计算机或其他机器上）查询，询问
10. pharmaceutical	[ˌfɑːmə'sjuːtɪkəl]	n.	药物
11. interactive	[ˌɪntər'æktɪv]	adj.	合作的；交互式的；人机对话的；互动的

Notes　注释

1. visual feedback　　　　　　　视觉反馈
2. head-mounted VR　　　　　　头戴式虚拟现实
3. additive manufacturing　　　　增材制造
4. virtual reality　　　　　　　　虚拟现实

5. simulated environment　　　　　　　仿真环境
6. 3D printing　　　　　　　　　　　　3D 打印
7. Radio Frequency Identification（RFID）　射频识别

Reference Translation　参考译文

1. 虚拟现实（VR）

虚拟现实（VR）是一种在模拟环境中由计算机生成的交互式体验。它主要包括听觉和视觉反馈，但也可能包含其他类型的感官反馈。

虚拟现实技术具有以下三个特征。

（1）沉浸性　指利用三维立体图像，给人一种身临其境的感觉。

（2）交互性　指利用一些传感设备进行交互，使用户感觉就像在真实客观世界中一样。

（3）想象性　指使用户沉浸其中并提高感性和理性认识，进而产生认知上的新意和构想。

一般的虚拟现实系统主要包括5个部分：专业图形处理计算机、应用软件系统、输入设备、输出设备和数据库。

2. 3D 打印

3D 打印是一种在计算机控制下连接或固化材料以创建三维物体的过程。图 3-12 展示了 3D 打印技术的工作原理，首先通过一组平行平面去截取零件的数字三维 CAD 模型，得到一系列足够薄的切片（厚度一般为 0.01~0.1mm），然后按照一定规则堆积起来即可得到整个零件。

如今，随着精度、可重复性和材料范围的增加，3D 打印被视为一种工业生产技术，其正式名称是增材制造。3D 打印对象可以具有非常复杂的形状或几何结构。

3. 射频识别（RFID）

射频识别（RFID）是一种自动识别和数据捕获的方法。RFID 利用电磁场来自动识别和跟踪附着在物体上的标签，这些标签包含电子存储的信息。

RFID 系统因应用不同其组成会有所不同，但基本都由电子标签、读卡器和数据管理系统三大部分组成。

（1）电子标签　电子标签具有智能读写和加密通信的功能，它是通过无线电波与读写设备进行数据交换的，其工作能量是由读卡器发出的射频脉冲提供的。

（2）读卡器　读卡器可将主机的读写命令通过天线传送到电子标签，再把从主机发往电子标签的数据加密，将电子标签返回的数据解密后送到主机。

（3）数据管理系统　数据管理系统完成数据信息的存储及管理，对卡进行读写控制等功能。

Unit 4 Mechanical Elements and Mechanisms

机械元件和机构

Part 1 Kinematic Sketch of Mechanism
机构运动简图

1. Component and Mechanism

Component is a motion unit that makes up a machine, which can be a part or a rigid combination of several parts. For example, nut and bolt are components that contain a single part, the connecting rod is a combination of some parts, as shown in Figure 4-1 and Figure 4-2.

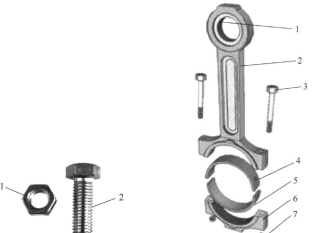

Figure 4-1 Nut and Bolt
1—Nut; 2—Bolt

Figure 4-2 Connecting Rod
1—Bushing 2—Connecting rod 3—Bolt
4—Upper bearing shell 5—Lower bearing shell
6—Bearing cap 7—Nut

Video 12

Mechanism is the basic unit of many mechanical equipment, is a combination of rigid or resistant bodies, so formed and connected that they move upon each other with definite relative motion. A mechanism can transmit or transform motion. A movable connection consisting of two components is called a kinematic pair.

2. Kinematic Sketch of Mechanism

In order to analyze the kinematic characteristics of a mechanism some simple lines and special symbols are usually used to describe the original components and kinematic pairs. This sketch, which can accurately describe the kinematic relationship of mechanism, is called the kinematic sketch of mechanism. The kinematic sketch only reflects the position of the components and the pairs at one moment. The kinematic sketches of commonly used mechanisms are shown in Figure 4-3, Figure 4-4, Figure 4-5 and Figure 4-6.

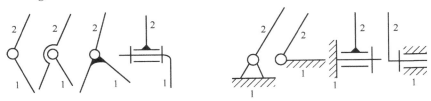

Figure 4-3 Kinematic Sketch of Revolute Pair

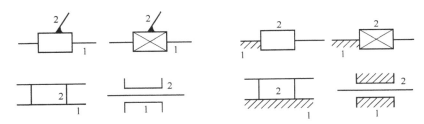

Figure 4-4 Kinematic Sketch of Sliding Pair

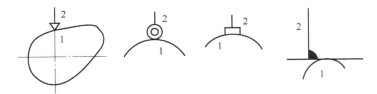

Figure 4-5 Kinematic Sketch of Cam Pair

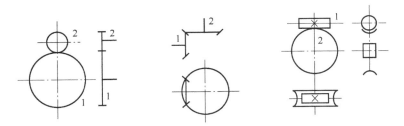

Figure 4-6 Kinematic Sketch of Gear Pair

Unit 4 Mechanical Elements and Mechanisms

机械元件和机构

3. Pipeline of Drawing a Kinematic Sketch of Mechanism

Taking Figure 4-7, the kinematic sketch of internal combustion engine as an example, the drawing pipeline of kinematic sketch of plane mechanism sketch is described as below.

Step 1: Analyze the mechanism, observe the relative motion of each component, and determine the frame, the original parts and the followers.

Step 2: Determine the type and number of components and kinematic pairs.

Step 3: Select the view plane which can fully reflect the motion characteristics of the mechanism.

Step 4: Determine the appropriate proportion.

Step 5: Starting with the original part, determine the relative position of each pair and draw the sketch with the prescribed symbols.

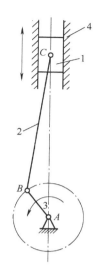

Figure 4-7 Kinematic Sketch of Internal Combustion Engine
1—Piston 2—Connecting rod 3—Crank 4—Frame

Vocabulary 词汇

1. component   [kəm'pəʊnənt]   n. 构件
2. mechanism   ['mekənɪzəm]   n. 机构
3. nut   [nʌt]   n. 螺母
4. bolt   [bəʊlt]   n. 螺栓
5. movable   ['muːvəbl]   adj. 可移动的；活动的
6. prescribe   [prɪ'skraɪb]   v. 规定；命令；指示
7. crank   [kræŋk]   n. 曲柄，曲轴
8. frame   [freɪm]   n. 构架，支架

Notes 注释

1. kinematic sketch of mechanism   机构运动简图
2. connecting rod   连杆
3. kinematic pair   运动副
4. kinematic characteristic   运动特性
5. revolute pair   转动副
6. sliding pair   移动副
7. cam pair   凸轮副
8. gear pair   齿轮副
9. internal combustion engine   内燃机

Reference Translation 参考译文

1. 构件与机构

构件是组成机器的运动单元,可以是一个零件或若干零件的刚性组合体。例如,螺母、螺栓是包含单个零件的构件,连杆是由连杆体、连杆头和螺栓、螺母等多个零件刚性组合而成的一个构件,如图4-1和图4-2所示。

机构是构成许多机械设备的基本单元,是由刚体或者有承载能力的物体连接而形成的组合体,它们在运动时彼此之间应该具有确定的相对运动。机构的作用是在刚体之间传递或转换运动。由两个构件直接接触而组成的可动的连接称为运动副。

2. 机构运动简图

为了对机构进行运动特性分析,通常采用一些简单的线条和特定的符号描绘原构件和运动副。这种能准确表达机构运动关系的简图称为机构运动简图。运动简图只能反映某一瞬时各构件和运动副间的位置关系。常用机构的简图符号如图4-3~图4-6所示。

3. 机构运动简图绘制示例

以图4-7所示内燃机运动简图为例,平面机构运动简图的绘制方法如下。

第1步:分析机构,观察各构件间的相对运动,确定机架、原动件和从动件。

第2步:确定构件和运动副的类型、数目。

第3步:选择能充分反应机构运动特性的视图平面。

第4步:确定适当的比例。

第5步:从原动件开始,确定各运动副的相对位置并用规定的符号绘制其运动简图。

Part 2 Mechanical Transmission Mechanism
机械传动机构

Commonly used mechanism includes gear, cam, linkage, chain and belt.

Video 13

1. Gear

The transmission of rotary motion from one shaft to another occurs in nearly every machine one can imagine. Gears constitute one of the best of the various means available for transmitting this motion. A gear is a wheel with very accurately shaped teeth. These teeth mesh with the teeth of another gear, thus providing a positive-motion drive. Various types of gears have been developed for different purposes, such as spur gear, bevel gear, or worm gear, as shown in Figure 4-8.

2. Cam

A part on a machine which mechanically changes cylindrical motion to reciprocating motion. Since the cam outline can be given a wide variety of shapes, many different types of motion can be secured. The purpose of a cam is to transmit various kinds of motion to other parts of a machine.

Though each cam appears to be quite different from the other, all the cams work in a similar

Unit 4 Mechanical Elements and Mechanisms
机械元件和机构

a) Spur Gear   b) Bevel Gear   c) Worm Gear

Figure 4-8 Various Types of Gears

way. In each case, as the cam is rotated or turned, another part in contact with the cam called follower, is moved either left and right, up and down, or in and out. The follower is usually connected to other parts on the machine to accomplish the desired action. Different types of cams are shown in Figure 4-9.

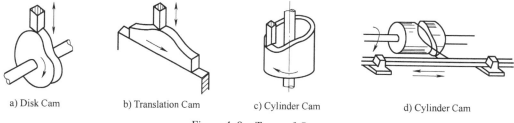

a) Disk Cam   b) Translation Cam   c) Cylinder Cam   d) Cylinder Cam

Figure 4-9 Types of Cams

3. Linkage

Linkage is a series of at least three rods or solid links that are connected by joints that permit the links to pivot. When one link is fixed the other links can move only in paths that are predetermined. Linkages are used to change the direction of motion, to transmit different kinds of motion or to provide variations in timing in different parts of a cycle by varying the lengths of the links in relation to each other.

The most commonly used planar linkage mechanism is a planar linkage mechanism consisting of four components, called a planar four-bar mechanism, as shown in Figure 4-10. According to the different motion forms of the two connecting rods, planar four-bar mechanism can be divided into three basic forms: double crank mechanism(Figure 4-11), crank-rocker mechanism(Figure 4-12) and double rocker mechanism(Figure 4-13).

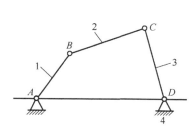   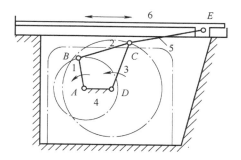

Figure 4-10 Planar Four-Bar Mechanism   Figure 4-11 Double Crank Mechanism

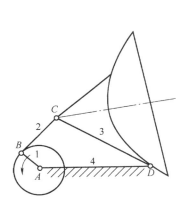
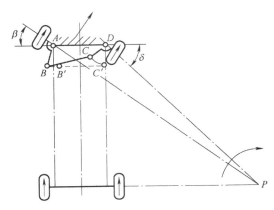

Figure 4-12　Crank-Rocker Mechanism

Figure 4-13　Double Rocker Mechanism

4. Chain Drives

A chain is a power transmission element made as a series of pin-connected links. The design provides for flexibility while enabling the chain to transmit large tensile forces When transmitting power between rotating shafts, the chain engages mating toothed wheels, called sprockets. Figure 4-14 shows a typical chain drive. The most common type of chain is the roller chain, in which the roller on each pin provides exceptionally low friction between the chain and the sprockets.

5. Belt Drives

A belt is a flexible power transmission element that seats on a set of pulleys or sheaves, as show in Figure 4-15.

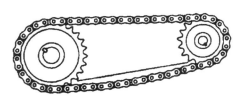
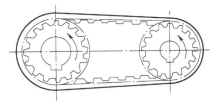

Figure 4-14　Roller Chain

Figure 4-15　Belt Drive

The belt is installed around the two sheaves while the center distance between them is reduced. Then the sheaves are moved apart, placing the belt in a rather high initial tension. When the belt is transmitting power, friction causes the belt to grip the driving sheave, increasing the tension on one side, called the "tight side". The tensile force in the belt exerts a tangential force on the driven sheave, and thus a torque is applied to the driven shaft.

Many types of belts are available, such as flat belts, V belts, circular belts, multi-wedge belt and others. See Figure 4-16 for examples.

Unit 4　Mechanical Elements and Mechanisms
机械元件和机构

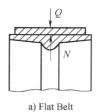 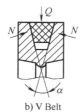 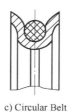 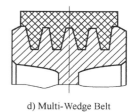

　　a) Flat Belt　　　　b) V Belt　　　　c) Circular Belt　　　d) Multi-Wedge Belt

Figure 4-16　Different Types of Belt

Vocabulary　词汇

1. linkage　　　　　['lɪŋkɪdʒ]　　　　n. 连杆
2. chain　　　　　[tʃeɪn]　　　　　n. 链子；链条
3. belt　　　　　　[belt]　　　　　n. 传动带；传送带
4. shaft　　　　　[ʃɑːft]　　　　　n. 轴
5. follower　　　　['fɒləʊə(r)]　　　n. 从动轮
6. pivot　　　　　['pɪvət]　　　　v. （使）在枢轴上旋转（或转动）
7. sprocket　　　['sprɒkɪt]　　　n. 链轮；链轮齿；链齿
8. pulley　　　　　['pʊli]　　　　　n. 滑轮；滑轮组
9. sheave　　　　[ʃiːv]　　　　　n. 滑轮；带轮

Notes　注释

1. mesh with　　　　　　　　与……啮合
2. spur gear　　　　　　　　直齿轮
3. bevel gear　　　　　　　　伞齿轮
4. worm gear　　　　　　　　蜗轮
5. cylindrical motion　　　　　圆柱运动
6. reciprocating motion　　　　往复运动
7. disk cam　　　　　　　　　盘形凸轮
8. translation cam　　　　　　平动凸轮
9. cylinder cam　　　　　　　圆柱凸轮
10. planar linkage mechanism　　平面连杆机构
11. planar four-bar mechanism　　平面四杆机构
12. double crank mechanism　　双曲柄机构
13. crank-rocker mechanism　　曲柄摇杆机构
14. double rocker mechanism　　双摇杆机构
15. tensile force　　　　　　　拉力
16. roller chain　　　　　　　滚子链
17. belt drive　　　　　　　　带传动
18. circular belt　　　　　　　圆形带
19. multi-wedge belt　　　　　多楔带

Reference Translation 参考译文

常用传动机构包括齿轮、凸轮、连杆、链条和传动带。

1. 齿轮

齿轮将旋转运动从一根轴传递到另一根轴,几乎存在于每一种机器上。齿轮是传递这种运动最有效的方法之一。齿轮的齿形非常精确。一只齿轮上的齿与另一只齿轮上的齿相啮合,这样就可产生强制传动。目前,已研制出各种齿轮装置以适应不同的用途,如正齿轮、伞齿轮或蜗轮,如图 4-8 所示。

2. 凸轮

凸轮作为机器上的一个零件,能机械地将圆周运动转变为往复运动。凸轮的功能是将各种形式的运动传递给机器中的其他零件。由于凸轮的外轮廓线可以是各种各样的形状,因而可获得许多不同形式的运动。

不同外形的凸轮工作原理都是相似的。在任何情况下,凸轮转动时,另一个与凸轮相接触的叫作从动件的零件就做左右、上下或内外运动。从动件通常与机器的其他零件相连接,以完成所要求的动作。不同类型的凸轮如图 4-9 所示。

3. 连杆

连杆是由至少三根杆或实心连杆组成的一系列通过允许连杆转动的接头链接。当一个连杆固定时,其他连杆只能在预定的路径中移动。与凸轮一样,连杆也用来改变运动方向,传递不同类型的运动,或者通过改变连杆之间的相对长度来改变一个循环中不同部分的时间。

平面连杆机构是由转动副将构件连接起来所组成的平面机构。它的主要作用是实现运动形式的变换。最常用的平面连杆机构是由四个构件组成的平面连杆机构,称为平面铰链四杆机构,如图 4-10 所示。平面铰链四杆机构根据其两连架杆的运动形式不同,可分为双曲柄机构(图 4-11)、曲柄摇杆机构(图 4-12)和双摇杆机构(图 4-13)三种基本形式。

4. 链条

链条是一种动力传动元件,由一系列销连接而成。这种设计提供了灵活性,同时使链条在转动轴之间传递动力时能够传递较大的拉力,链条与称为链轮的啮合齿轮啮合。图 4-14 所示为典型的链传动。最常见的链条类型是滚子链,其中每个销上的滚子在链条和链轮之间的摩擦极低。

5. 传动带

传动带是一种灵活的动力传动元件,固定在一组带轮上,如图 4-15 所示。传动带安装在两个带轮周围,同时它们之间的中心距离减小。然后,将带轮分开,使传动带处于相当高的初始张力。当传动带传递动力时,摩擦使传动带抓住主动带轮,增加了一侧的张力,称为"紧侧"。传动带中的拉力对从动带轮施加切向力,从而对从动轴施加扭矩。有许多类型的传动带可供选择,如平带、V 带、圆形带、多楔带等。

Part 3 Mechanical Connection Components
机械连接部件

1. Shafts

Virtually all machines contain shafts. The most common shape for shafts is circular and the cross section can be either solid or hollow (hollow shafts can result in weight savings). Rectangular shafts are sometimes used, as in screwdriver blades, socket wrenches and control knob stems.

Classification of shafts

1) Based on the load, shafts can be divided into different classes.

① Rotary shaft supports transmission part and transfers power (Figure 4-17). It bears both torque and bending moment.

Video 14

② Spindle supports rotating parts and does not transmit power (Figure 4-18, Figure 4-19). It only bears bending moment. The spindle is divided into fixed spindle (the shaft does not rotate while working, Figure 4-18) and rotating spindle (the shaft rotates while working, Figure 4-19).

③ Transmission shaft mainly plays the role of transmitting power (Figure 4-20). It mainly bears the torque.

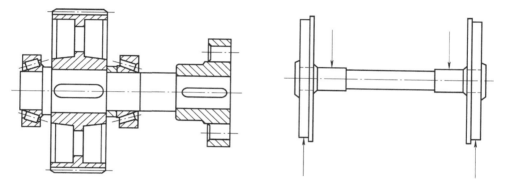

Figure 4-17 Rotary Shaft   Figure 4-18 Spindle Shaft of Trains

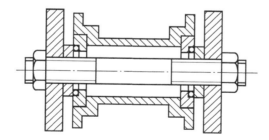

Figure 4-19 Spindle Shaft of Front Wheel of Bicycle

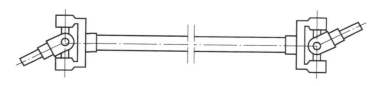

Figure 4-20 Transmission Shaft

2) Based on the structure, shafts can be divided into smooth shafts(Figure 4-21) and stepped shafts(Figure 4-22).

3) Based on the geometric shape, shafts can be divided into straight shafts(Figure 4-23), crank shafts(Figure 4-24) and flexible shafts(Figure 4-25).

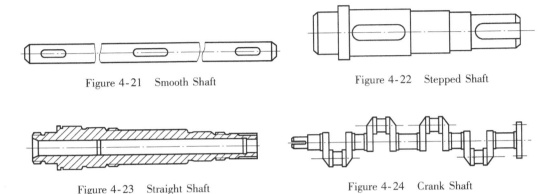

Figure 4-21　Smooth Shaft　　　　　　　Figure 4-22　Stepped Shaft

Figure 4-23　Straight Shaft　　　　　　　Figure 4-24　Crank Shaft

2. Couplings

A coupling is a device for connecting the ends of adjacent shafts. In machine construction, couplings are used as a semipermanent connection between adjacent rotating shafts. The connection is permanent in the sense that it is not meant to be broken during the useful life of the machine, but it can be broken and restored in an emergency or when worn parts are replaced.

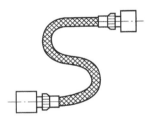

Figure 4-25　Flexible Shaft

There are several types of shaft couplings, their characteristics depend on the purpose for which they are used. If an exceptionally long shaft is required in a manufacturing plant or a propeller shaft on a ship, it is made in sections that are coupled together with rigid couplings. Examples of rigid couplings are shown in Figure 4-26a and Figure 4-26b.

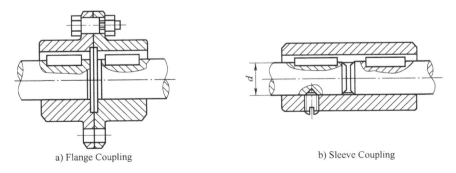

a) Flange Coupling　　　　　　　b) Sleeve Coupling

Figure 4-26　Rigid Couplings

In connecting shafts belonging to separate devices(such as an electric motor and a gearbox), precise aligning of the shafts is difficult and a flexible coupling is used. Examples of flexible couplings are shown in Figure 4-27a and Figure 4-27b. This coupling connects the shafts in such a way as to minimize the harmful effects of shaft misalignment. Flexible couplings can also serve to reduce the intensity of shock loads and vibrations transmitted from one shaft to another.

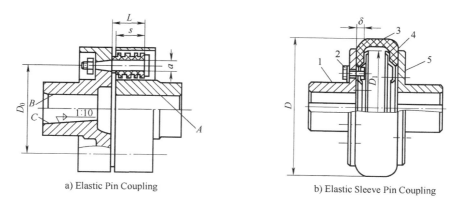

a) Elastic Pin Coupling   b) Elastic Sleeve Pin Coupling

Figure 4-27 Flexible Couplings

3. Keys

A key is the machinery component placed at the interface between a shaft and the hub of a power-transmitting element for the purpose of transmitting torque. The key is demountable to facilitate assembly and disassembly of the shaft system. It is installed in an axial groove machined into the shaft, called a keyseat. A similar groove in the hub of the power-transmitting element is usually called a keyway, but it is more properly also a keyseat. The key is typically installed into the shaft keyseat first; then the hub keyseat is aligned with the key, and the hub is slid into position.

The keys are mainly divided into the following categories: flat keys(Figure 4-28), semi-circular keys(Figure 4-29), splines(Figure 4-30), wedge keys, tangential keys and so on.

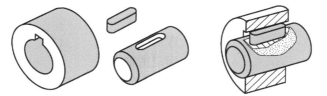

Figure 4-28 Flat Keys

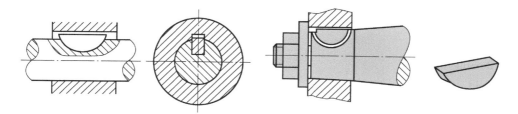

Figure 4-29 Semi-Circular Keys

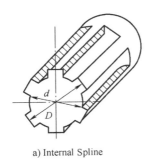
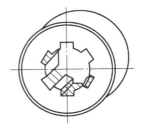

a) Internal Spline   b) External Spline

Figure 4-30 Splines

Vocabulary 词汇

1. hollow   [ˈhɒləʊ]   adj. 中空的；空心的
2. spindle   [ˈspɪndl]   n. 主轴
3. coupling   [ˈkʌplɪŋ]   n. 联轴器
4. propeller   [prəˈpelə(r)]   n. 螺旋桨
5. key   [kiː]   n. 键
6. keyseat   [ˈkiːsit]   n. 键槽
7. hub   [hʌb]   n. 轮毂
8. spline   [splaɪn]   n. 花键

Notes 注释

1. screwdriver blade   螺钉旋具刀片
2. socket wrench   套筒扳手
3. control knob   控制旋钮
4. rotary shaft   转动轴
5. bending moment   弯矩
6. fixed spindle   固定主轴
7. rotating spindle   旋转主轴
8. transmission shaft   传动轴
9. smooth shaft   光轴
10. stepped shaft   阶梯轴
11. straight shaft   直轴
12. crank shaft   曲轴
13. flexible shaft   挠性轴
14. rigid coupling   刚性联轴器
15. flange coupling   法兰联轴器
16. sleeve coupling   套筒联轴器
17. flexible coupling   挠性联轴节

Unit 4 Mechanical Elements and Mechanisms

机械元件和机构

18. elastic pin coupling	弹性柱销联轴器
19. elastic sleeve pin coupling	弹性套柱销联轴器
20. axial groove	轴向槽
21. flat key	平键
22. semi-circular key	半圆键
23. wedge key	楔形键
24. tangential key	切向键
25. internal spline	内花键
26. external spline	外花键

Reference Translation 参考译文

1. 轴

实际上，几乎所有的机器都装有轴。轴最常见的形状是圆形，其截面可以是实心的，也可以是空心的。有时也采用矩形轴，例如螺钉旋具刀片、套筒扳手和控制旋钮的杆。

轴的分类如下：

1）按轴受载情况，轴可分为不同类型。

① 旋转轴支承传动零件并传递动力（见图 4-17），它同时承受转矩和弯矩。

② 主轴只支承回转零件而不传递动力，它只承受弯矩（见图 4-18、图 4-19）。主轴又分为固定主轴（工作时轴不转动，见图 4-18）和转动主轴（工作时轴转动，见图 4-19）。

③ 传动轴主要起传递动力的作用，即主要承受转矩。

2）按轴的结构形状，轴可分为光轴（见图 4-21）和阶梯轴（见图 4-22）。

3）按几何轴线形状，轴可分为直轴（见图 4-23）、曲轴（见图 4-24）和挠性轴（见图 4-25）。

2. 联轴器

联轴器是用来把两个相邻轴端连接起来的装置。在机械结构中，联轴器被用来实现相邻的两根旋转轴之间的半永久性连接。在机器的正常使用期间内，这种连接一般不必拆开。但是在紧急情况下或者需要更换已磨损的零部件时，可以先把联轴器拆开，然后再接上。

联轴器有几种类型，它们的特性随其用途而定。如果制造厂或者船舶的螺旋桨需要一根特别长的轴，可以采用分段的方式将其制造出来，然后采用刚性联轴器将各段连接起来。刚性联轴器的例子如图 4-26a、b 所示。

在把属于不同的设备（例如一个电动机和一个变速器）的轴联接起来的候，要把这些轴精确地对准是比较困难的，此时可以采用弹性联轴器。弹性联轴器的例子如图 4-27a、b 所示。这种联轴器联接轴的方式可以把由于被联接轴之间的轴线不重合所造成的有害影响减少到最低程度。弹性联轴器也可以用来减轻从一根轴传递到另一根轴上的冲击载荷和振动的强度。

3. 键

键是放置在轴和动力传输元件轮毂之间的接口处的机械部件，用于传输扭矩。键是可拆卸的，便于轴系统的组装和拆卸。它被安装在轴上加工成的轴向槽中，称为键槽。在动力传输元件轮毂上，类似的凹槽通常被称为键沟，但更恰当的说法是键槽。通常先将键安装到轴键槽中，然后将轮毂键槽与键对齐，然后将轮毂滑入到位。

键主要分为以下几类，平键（见图 4-28）、半圆键（见图 4-29）、花键（见图 4-30）、楔键和切向键等。

Part 4 Manufacturing Process
制造工艺

1. Basic Machining Techniques

(1) Turning　Conventional lathe operations include cylindrical turning, facing, groove cutting, boring, internal turning, taper turning and thread cutting. A turning example is shown in Figure 4-31.

Video 15

(2) Milling　Milling is a machining process that is carried out by means of a multi-edge rotating tool known as a milling cutter. In this process, metal removal is achieved through combining the rotary motion of the milling cutter and linear motions of the workpiece simultaneously. Milling operations are employed in producing flat, contoured and helical surfaces as well as for thread and gear-cutting operations. A milling example is shown in Figure 4-32.

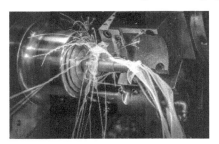

Figure 4-31　Turning Example

Figure 4-32　Milling Example

(3) Grinding　Grinding is a manufacturing process that involves the removal of metal by employing a rotating abrasive wheel. Generally, grinding is considered to be a finishing process that is usually used for obtaining high dimensional accuracy and better surface finish. Grinding can be performed on flat, cylindrical, or even internal surfaces by employing specialized machine tools, which are referred to as grinding machines. A grinding example is shown in Figure 4-33.

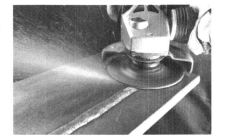

Figure 4-33　Grinding Example

2. Hot Working Processes

There are three basic methods of metal working: hot working, cold working, and extruding. The process chosen for a particular application depends upon the metal involved and the part required, in

Unit 4 Mechanical Elements and Mechanisms
机械元件和机构

some instances you might employ both hot and cold working methods in making a single part.

(1) Forming There are generally two types of hot forming process: rolling and forging. Example of forming is shown in Figure 4-34.

Rolling is a process whereby the shape of the hot metal is altered by the action of the rollers which acts to "squeeze" the hot metal into desired shape and thickness.

Forging is another important hot working method. It is used in producing components of all shapes and sizes, from quite small items to large units weighting several tons. In forging, the metal is pounded by hammers or squeezed between a pair of shaped dies. The die acts as a hammer that can "pound" the hot metal into shape. Forging is used primarily to produce a rough shape.

(2) Casting Casting is a manufacturing process in which molten metal is poured or injected and allowed to solidify in a suitably shaped mold cavity. During or after cooling, the cast part is removed from the mold and then processed for delivery. Examples of casting parts are shown in Figure 4-35.

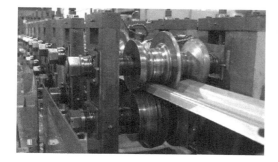

Figure 4-34 Forming Example

Figure 4-35 Examples of Casting Parts

Casting is mainly divided into three classes: sand casting, pressure die casting and investment casting.

1) Sand casting is used to make large parts. Molten metal is poured into a cavity formed out of compacted molding sand.

2) Pressure die casting makes possible the economical quantity production of intricate castings at a rapid rate. Such castings are characterized by high dimensional accuracy, good surface finish, and economy of metal: they require little or no final machining.

3) Investment casting is also known as the lost wax process. This process is one of the oldest manufacturing processes.

Vocabulary 词汇

1. turning   ['tɜːnɪŋ]   n. 车削
2. milling   ['mɪlɪŋ]   n. 铣削
3. grinding   ['graɪndɪŋ]   n. 磨削
4. helical   ['helɪkl]   adj. 螺旋的；螺旋形的
5. extruding   [ɪk'struːdɪŋ]   v. （被）挤压出
6. rolling   ['rəʊlɪŋ]   n. 滚压

7. forging	[ˈfɔːdʒɪŋ]	n. 锻造
8. roller	[ˈrəʊlə(r)]	n. 滚筒；滚轴
9. pound	[paʊnd]	v. 反复击打；连续砰砰地猛击
10. die	[daɪ]	n. 模具
11. casting	[ˈkɑːstɪŋ]	n. 铸造，铸件
12. pour	[pɔː(r)]	v. 使（液体）连续流出；倾倒；倒出
13. solidify	[səˈlɪdɪfaɪ]	v. （使）凝固，变硬
14. intricate	[ˈɪntrɪkət]	adj. 错综复杂的

Notes 注释

1. cylindrical turning	车外圆
2. groove cutting	切槽
3. internal turning	镗孔
4. taper turning	车锥度
5. thread cutting	车螺纹
6. milling cutter	铣刀
7. gear-cutting	齿轮切削
8. rotating abrasive wheel	旋转砂轮
9. hot working	热加工
10. cold working	冷加工
11. forming process	成形过程
12. shaped mold cavity	成形模腔
13. cast part	铸件
14. sand casting	砂型铸造
15. pressure die casting	压力铸造
16. investment casting	熔模铸造
17. lost wax process	失蜡铸造

Reference Translation 参考译文

1. 基本加工技术

（1）车削 普通车床的基本操作包括车外圆、车端面、车槽、镗孔、车内圆、锥度车削和车螺纹。

（2）铣削 铣削是通过一种称为铣刀的多刃旋转刀具完成机加工的工艺。在这个过程中，金属的切除是通过刀具的旋转和同时相对于工件的线性移动的组合完成的。铣削操作用于平面、轮廓、螺旋面以及螺纹和齿轮的加工。

（3）磨削 磨削是一种使用旋转的砂轮去除金属的加工的工艺。通常磨削被认为是一种精加工工艺，用于获得高尺寸精度和更好的表面质量。磨削适用于平面加工、外圆柱面加工，甚至是内表面加工，这些加工是通过一种叫作磨床的专用机床来完成的。

Unit 4　Mechanical Elements and Mechanisms
机械元件和机构

2. 金属的热加工

金属加工有三种基本方法：热加工、冷加工和挤压。在加工某一个工件的时候，既可以用热加工也可以用冷加工，应该根据所采用的金属和对零件的要求来选择加工方法。

（1）成形　热成形通常有两种方法：轧制和锻造。轧制即热的金属在轧制机上被挤压成需要的形状和厚度。轧制的一个好处是可以细化金属晶粒，而细化后的晶粒可以得到更好的物理特性。

锻造是另外一种重要的热加工方法。锻造通常用于生产各种尺寸和形状的零件，从很小的零件到几吨重的零件都有。锻造时金属被锤连续撞击或者被一副成形模具挤压成形。模具就像锤子一样撞击金属使之成形。锻造主要用于粗加工。

（2）铸造　铸造是将熔融的金属浇灌或注射到合适形状的型腔中固化的加工工艺。在冷却过程中或冷却以后，铸件从模型中取出并交付使用。铸件示例如图4-35所示。

铸造加工主要分为三类：砂型铸造、压力铸造和熔模铸造。

1）砂型铸造用于生产大型零件。将熔化的金属倒入由型砂做成的铸型中。

2）压力铸造使快速、经济大量地生产复杂铸件成为可能。这些铸件的特点是尺寸精度高，表面粗糙度小，节约材料，几乎不用精加工。

3）熔模铸造也称为失蜡铸造。这是最古老的制造工艺之一。

Unit 5 Electrical and Electronic Technology

电工电子技术

Part 1 Electrical Foundation
电工基础

Video 16

Electronic circuits are integral parts of nearly all the technological advances being made in our lives today. Television, radio, phones and computers immediately may come to mind, but electronics are also used in automobiles, kitchen appliances, medical equipment and industrial controls. At the heart of these devices are active components, or components of the circuit that electronically control electron flow, like semiconductors. Unlike active components, passive components, such as resistors, capacitors and inductors, can't control the electron flow with electronic signals.

1. Resistor

As its name implies, a resistor is an electronic component that resists the flow of electric current in a circuit. Figure 5-1 is a resistor with a variety of resistance values.

Resistors are generally classified as either fixed or variable.

1) Fixed-value resistors are simple passive components that always have the same resistance within their prescribed current and voltage limits. They are available in a wide range of resistance values, from less than 1 ohm to several million ohms.

2) Variable resistors are simple passive components, such as volume controls and dimmer switches, which change the effective length or effective temperature of a resistor when you turn a knob or move a slide control.

Unit 5 Electrical and Electronic Technology

电工电子技术

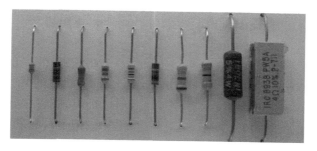

Figure 5-1 Resistors with Different Resistance Values

2. Inductor

An inductor is an electronic component consisting of a coil of wire with an electric current running through it, creating a magnetic field. The model is shown in Figure 5-2.

The unit for inductance is the henry(H), named after Joseph Henry, an American physicist who discovered inductance. One henry is the amount of inductance that is required to induce 1 volt of electromotive force when the current is changing at 1 ampere per second.

3. Capacitor

Capacitance is the ability of a device to store electric charge, and the electronic component that stores electric charge is called a capacitor, as shown in Figure 5-3.

Figure 5-2 Inductor Model

The capacitance of a capacitor is the amount of charge it can store per unit of voltage. The unit for measuring capacitance is the farad(F), named for Faraday, and is defined as the capacity to store 1 coulomb of charge with an applied potential of 1 volt. One coulomb(C) is the amount of charge transferred by a current of 1 ampere in 1 second.

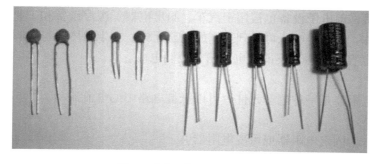

Figure 5-3 Capacitor Model

Vocabulary   词汇

1. electrical   [ɪˈlektrɪkl]   adj. 电的；发电的
2. electronic   [ˌɪlekˈtrɒnɪk]   adj. 电子的；电子元器件的

3. immediately	[ɪˈmiːdiətli]	adv. 立即；马上；即刻；直接地
4. resistor	[rɪˈzɪstə(r)]	n. 电阻器
5. classify	[ˈklæsɪfaɪ]	v. 将……分类，将……归类；划分
6. ohm	[əʊm]	n. 欧姆（电阻单位）
7. electromechanical	[ɪˌlɛktrəʊmɪˈkænɪk(ə)l]	adj. 机电的
8. inductor	[ɪnˈdʌktə]	n. 电感器；电感线圈；引导者
9. knob	[nɒb]	n. 把手；旋钮
10. inductance	[ɪnˈdʌktəns]	n. 电感
11. henry	[ˈhenri]	n. 亨利（电感单位）
12. ampere	[ˈæmpeə]	n. 安培（电流单位）
13. capacitor	[kəˈpæsɪtə(r)]	n. 电容器
14. capacitance	[kəˈpæsɪtəns]	n. 电容，电容量
15. voltage	[ˈvəʊltɪdʒ]	n. 电压；伏特
16. faraday	[ˈfærədeɪ]	n. 法拉第（电量单位）
17. coulomb	[ˈkuːlɒm]	n. 库伦（电量单位）
18. potential	[pəˈtenʃl]	n. 电位；电势；电压

Notes 注释

1. active component　　有源元件
2. passive component　　无源元件
3. magnetic field　　磁场
4. electric charge　　电荷
5. electronic component　　电子元件

Reference Translation 参考译文

如今，电子电路几乎包含在我们生活中所有新产生的技术中。我们也许会想到电视、广播、电话和计算机，但电子设备也被用于汽车、厨房电器、医疗设备和工业控制。这些器件的核心是有源元件，或者是电流可受到控制的电路元件，如半导体。与有源元件不同，无源元件（如电阻、电容和电感）不能用电子信号控制电流。

1. 电阻

顾名思义，电阻器是一种电子元件，可以抵抗电路中的电流流动。图 5-1 所示为具有各种电阻值的电阻器。

电阻器通常分为固定电阻器或可变电阻器。

1）固定电阻器是简单的无源元件，在规定的电流和电压限制范围内始终具有相同的电阻值。它们的电阻值范围很广，从小于 1 欧姆到几百万欧姆不等。

2）可变电阻器是简单的机电设备，例如音量控制器和调光器开关，当转动旋钮或移动滑动控制器时，它们会改变电阻器的有效长度或有效温度。

2. 电感器

电感器是一种由线圈组成的电子元件，电流流过线圈而形成磁场。其模型如图 5-2

所示。

电感的单位是亨利（H），以发现电感的美国物理学家约瑟夫·亨利来命名。1H 是当电流以 1A/s 的速度变化时，感应 1V 电动势所需的电感量。

3. 电容器

电容是设备存储电荷的能力，存储电荷的电子元件称为电容器，如图 5-3 所示。

电容器的电容是每单位电压可以存储的电荷量。测量电容的单位是法拉第（F），以法拉第来命名。一个电容器，如果带 1C 的电量时两极间的电势差是 1V，这个电容器的电容就是 1F。1C 是在 1s 内通过 1A 的电流所传输的电荷量。

Part 2 Electronic Components
电子元器件

An electronic component is a basic discrete device in an electronic system used to affect electrons or their associated fields. Electronic components are mostly industrial products, available in a singular form. Some example electronic components are shown in the following figure.

Figure 5-4 Types of Electronic Components

Video 17

Electronic components have a number of electrical terminals or leads. These leads connect to create an electronic circuit with a particular function. Basic electronic components may be packaged discretely, or integrated inside of packages such as semiconductor integrated circuits.

1. Types of Electronic Components

Electronic Components are of two types: Active and Passive Electronic Components.

1）Passive electronic components are those that do not have gain or directionality, for example, resistors, capacitors, diodes, and inductors.

2）Active electronic components are those that have gain or directionality, for example transistors, integrated circuits, logic gates.

2. Diode

A Diode is an electronic device that allows current to flow in one direction only. It is a semiconductor that consists of a PN junction. Both P-type and N-type silicon will conduct electricity just like any conductor; however, if a piece of silicon is doped P-type in one section and N-type in an adjacent section, current will flow in only one direction across the junction between the two regions. Diode is one of the most basic semiconductor devices.

3. Transistor

Transistor is one of the basic components of semiconductor, which has the function of current amplification. Transistor contains two very close PN junctions on a semiconductor substrate. Two PN junctions divide the whole semiconductor into three parts, the middle part is the base area, the two sides are the launch zone and the collector area. Based on the arrangement, there are two types of transistor, PNP and NPN.

4. Amplifiers

Amplifiers are generic terms used to describe circuits that amplify their input signals. Amplifier are classified according to their circuit configuration and operating mode. There are many forms of electronic circuits that are classified as amplifiers, from OP amplifiers and small signal amplifiers to large signal amplifiers and power amplifiers.

5. Relays

A relay is a special control component that acts like an active contact. When the current passes through a fixed value, the contact is disconnected(or connected), and the current interrupt(or renew) to control components in the same circuit or other circuit.

There are different types of relays including electromagnetic relays, latching relays, electronic relays, non-latching relays, multi-dimensional relays and thermal relays.

Vocabulary 词汇

1. associated	[əˈsəʊsieɪtɪd]	adj. 有关联的，相关的，有联系的
2. singular	[ˈsɪŋgjələ(r)]	n. 单数 adj. 单数的，单一的
3. terminal	[ˈtɜːmɪnl]	n. 终点站；终端，终端机
4. particular	[pəˈtɪkjələ(r)]	adj. 专指的，特指的；特别的，特定的
5. gain	[geɪn]	n. 增益
6. directionality	[dɪrekʃənˈælɪti]	n. 方向性；定向性
7. diode	[ˈdaɪəʊd]	n. 二极管
8. transistor	[trænˈzɪstə(r)]	n. 晶体管
9. substrate	[ˈsʌbstreɪt]	n. 底层，基底，基层
10. amplifier	[ˈæmplɪfaɪə(r)]	n. 放大器；扩音器，扬声器
11. relay	[ˈriːleɪ]	n. 继电器
12. configuration	[kənˌfɪgəˈreɪʃn]	n. 配置
13. electromagnetic	[ɪˌlektrəʊmægˈnetɪk]	adj. 电磁的

Notes 注释

1. integrated circuit   集成电路

2. logic gate 逻辑门
3. PN junction PN 结
4. launch zone 发射区
5. collector area 集电极区
6. generic term 通用术语
7. OP amplifier 运算放大器
8. thermal relay 热继电器
9. multi-dimensional 多维的，多维度的

Reference Translation 参考译文

电子元器件是电子系统中的基本分立元器件，用来影响电子或电磁场。电子元器件大多应用于工业产品，以单一形式提供。图 5-4 给出了一些电子元器件的示例。

电子元器件有许多端子或引线。这些导线连接在一起，形成具有特定功能的电子电路。基本的电子元器件可以独立封装，或者集成在诸如半导体集成电路内。

1. 电子元器件的分类

电子元器件有两种类型：无源和有源。

1）无源电子元器件是指那些没有增益或方向性的元器件，例如电阻、电容、二极管和电感器。

2）有源电子元器件是具有增益或方向性的元器件，例如晶体管、集成电路、逻辑门。

2. 二极管

二极管是一种电子装置，它只允许电流向一个方向流动。它是一种由 PN 结组成的半导体。P 型和 N 型硅会像任何导体一样导电；但是，如果一块硅在一个截面上掺杂 P 型半导体，而在相邻截面上掺杂 N 型半导体，那么电流只会沿一个方向流过两个区域之间的结。二极管是最基本的半导体器件之一。

3. 晶体管

晶体管是一种基本的半导体器体，具有电流放大功能。晶体管在半导体基底上有两个非常紧密的 PN 结。两个 PN 结将整个半导体分成三部分，中间部分是基极区，两侧是发射极区和集电极区。根据排列方式不同，晶体管分为两种类型，即 PNP 型和 NPN 型。

4. 放大器

放大器是用来描述能放大输入信号的电路的通用术语。放大器根据它们的电路结构和工作模式进行分类。有许多形式的电子电路被归为放大器，从运算放大器、小信号放大器到大信号放大器和功率放大器。

5. 继电器

继电器是一种特殊的控制元件，其作用类似于主动触点。当电流超过某个固定值时，触点断开（或连接），电流中断（或续通），以此来控制同一电路中或其他电路中的元件。继电器的类型有电磁继电器、闭锁继电器、电子继电器、非闭锁继电器、多维继电器和热继电器。

Part 3 Kirchhoff's Law
基尔霍夫定律

Kirchhoff's laws are basic laws of voltage and current in electrical circuits. They were first described in 1845 by German physicist Gustav Kirchhoff. These two rules are commonly known as: Kirchhoff's Current Law(KCL) which describes the basic law of the current flowing around a closed circuit; and Kirchhoff's Voltage Law(KVL) which describes the basic law of the voltage in a closed circuit. These laws can be applied in time and frequency domains.

1. Kirchhoff's Current Law(KCL)

The current entering any junction is equal to the current leaving that junction. That means, in Figure 5-5,

Figure 5-5 Kirchhoff's Current Law

$$i_2 + i_3 = i_1 + i_4$$

This law is also called Kirchhoff's First Law, or Kirchhoff's Junction Rule. This law states that, for any node(junction) in an electrical circuit, the sum of currents flowing into that node is equal to the sum of currents flowing out of that node; or equivalently: The algebraic sum of currents in a network of conductors meeting at a point is zero. Recalling that current is a signed(positive or negative) quantity reflecting direction towards or away from a node, this principle can be succinctly stated as:

$$\sum_{k=1}^{n} I_k = 0$$

where n is the total number of branches with currents flowing towards or away from the node.

The law is based on the conservation of charge where the charge(measured in coulombs) is the product of the current(in amperes) and the time(in seconds). If the net charge in a region is constant, the KCL will hold on the boundaries of the region. This means that KCL relies on the fact that the net charge in the wires and components is constant.

2. Kirchhoff's Voltage Law(KVL)

The sum of all the voltages around a loop is equal to zero. That means, in Figure 5-6,

$$u_1 + u_2 + u_3 - u_4 = 0$$

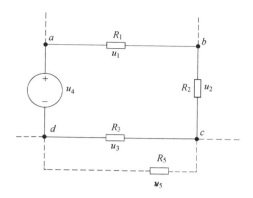

Figure 5-6　Kirchhoff's Voltage Law

This law is also called Kirchhoff's Second Law, Kirchhoff's Loop(or Mesh) Rule.

This law states that the directed sum of the potential differences(voltages) around any closed loop is zero. Similarly to KCL, it can be stated as:

$$\sum_{k=1}^{n} U_k = 0$$

where n is the total number of voltages measure.

Vocabulary　词汇

1. commonly　　　　['kɒmənli]　　　　adv. 通常，常常
2. domain　　　　　[də'meɪn]　　　　 n. 领域，范围；定义域
3. junction　　　　 ['dʒʌŋkʃn]　　　　n. 连接，接合；交叉点，接合点
4. state　　　　　　[steɪt]　　　　　 v. 陈述，说明；规定
5. node　　　　　　[nəʊd]　　　　　 n. 节点
6. equivalently　　 [ɪ'kwɪvələntli]　　 adv. 等价地
7. algebraic　　　　[ˌældʒɪ'breɪɪk]　　adj. 代数的
8. positive　　　　　['pɒzətɪv]　　　　n. 正数
9. negative　　　　['negətɪv]　　　　 n. 负数
10. succinctly　　　[sək'sɪŋktli]　　　 adv. 简洁地，简明地
11. boundary　　　 ['baʊndri]　　　　n. 边界，界限
12. loop　　　　　　[luːp]　　　　　　n. 回路
13. mesh　　　　　 [meʃ]　　　　　　n. 网格
14. product　　　　 ['prɒdʌkt]　　　　n. 乘积

Notes　注释

1. Kirchhoff's Laws　　　　　基尔霍夫定律
2. be known as　　　　　　　被称为；以……著称
3. Kirchhoff's Current Law　　基尔霍夫电流定律

4. Kirchhoff's Voltage Law　　基尔霍夫电压定律
5. is equal to　　等于，等同于
6. net charge　　净电荷
7. Kirchhoff's Junction Rule　　基尔霍夫节点定律
8. Kirchhoff's Loop Rule　　基尔霍夫回路定律

Reference Translation　参考译文

基尔霍夫定律是电路中电压和电流所遵循的基本规律。它们最早是在1845年由德国物理学家古斯塔夫·基尔霍夫提出的。这两个规则通常被称为：基尔霍夫电流定律（KCL），它描述的是在一个闭合回路中电流所遵循的规律；基尔霍夫电压定律（KVL），它描述的是在一个闭合回路中电压所遵循的规律。这些规律可以应用于时域和频域。

1. 基尔霍夫电流定律（KCL）

进入任何节点的电流等于离开该节点的电流。这意味着，在图5-5中有：

$$i_2 + i_3 = i_1 + i_4$$

这条定律也被称为基尔霍夫第一定律，或基尔霍夫节点定律。该定律规定，对于电路中的任何节点，流入该节点的电流之和等于流出该节点的电流之和；或等效地，流过导体网络中一个点的电流的代数和为零。考虑到电流是一个有符号（正或负）的量，它反映了流向还是远离一个节点的方向，这个定律可以简单地表述为

$$\sum_{k=1}^{n} I_k = 0$$

其中n是电流流向或远离节点的分支总数。

该定律基于电荷守恒，其中电荷（以库仑为单位）是电流（以安培为单位）与时间（以秒为单位）的乘积。如果一个区域的净电荷是恒定的，KCL在该区域的边界上始终成立。这意味着KCL依赖于导线和元件中的净电荷是恒定的这一事实。

2. 基尔霍夫电压定律（KVL）

回路的电压之和等于零。这意味着，在图5-6中有：

$$u_1 + u_2 + u_3 - u_4 = 0$$

这条定律也被称为基尔霍夫第二定律，基尔霍夫环（网格）定律。

该定律表明，任何闭合回路的电位差（电压）的有向和为零。与KCL类似，它可以表述为

$$\sum_{k=1}^{n} U_k = 0$$

其中n是测量的电压总数。

Integrated Circuit
集成电路

1. Definition

An integrated circuit(also referred to as an IC, or a chip) is a set of electronic circuits on one small flat piece(or "chip") of semiconductor material that is normally silicon. An illustration of chips is shown in Figure 5-7.

a) Individual Integrated Circuits

b) Integrated Circuits Mounted on PCB (Printed Circuit Board)

Figure 5-7 Examples of Integrated Circuits

The integration of large numbers of tiny transistors into a small chip results in circuits that are smaller, cheaper, and faster than those constructed of discrete electronic components. The IC's mass production capability and reliability has ensured the rapid adoption of standardized ICs in place of designs using discrete transistors. ICs are now used in virtually all electronic equipment and have revolutionized the world of electronics.

Video 19

2. Development

Integrated circuits were made practical by mid-20th-century technology advancements in semiconductor device fabrication. Since their origins in the 1960s, the size, speed, and capacity of chips have progressed enormously. These advances roughly follow Moore's Law, which states that the number of transistors in a dense integrated circuit doubles approximately every two years.

In the early days of simple integrated circuits, each chip has only a few transistors, and the low degree of integration meant the design process was relatively simple. Manufacturing yields were also quite low by today's standards. As the technology progressed, millions, then billions of transistors could be placed on one chip, and good designs required thorough planning. Table 5-1 shows the different generations of ICs.

Table 5-1 Different Generations of ICs

Name	Scale	Year	Transistors number	Logic gates number
SSI	Small-scale integration	1964	1-10	1-12
MSI	Medium-scale integration	1968	10-500	13-99
LSI	Large-scale integration	1971	500-20,000	100-9,999
VLSI	Very large-scale integration	1980	20,000-1,000,000	10,000-99,999
ULSI	Ultra-large-scale integration	1984	1,000,000 and more	100,000 and more

3. Advantages and Disadvantages

ICs have two main advantages over discrete circuits: low cost and good performance.

1) Cost is low because the chips, with all their components, are printed as a unit by photolithography rather than being constructed one transistor at a time. Furthermore, packaged ICs use much less material than discrete circuits.

2) Performance is good because the IC's components switch quickly and consume comparatively little power because of their small size and close proximity.

The main disadvantage of ICs is the high cost to design them and fabricate the required photomasks. This high initial cost means ICs are only practical when high production volumes are anticipated.

Vocabulary 词汇

1. rapid   ['ræpɪd]   adj. 迅速的，急促的
2. fabrication   [ˌfæbrɪ'keɪʃn]   n. 制造，制作
3. enormously   [ɪ'nɔːməsli]   adv. 非常地，极其地
4. photolithography   [ˌfəʊtəlɪ'θɒɡrəfɪ]   n. 光刻
5. photomask   ['fəʊtəʊmɑːsk]   n. 光掩模
6. volume   ['vɒljuːm]   n. 体积；量
7. silicon   ['sɪlɪkən]   n. 硅

Notes 注释

1. Moore's law   摩尔定律
2. initial cost   初始成本
3. mass production   批量生产
4. electronic equipment   电子设备
5. tiny transistor   小晶体管
6. electronic circuit   电子电路
7. discrete electronic component   分立电子元器件
8. standardized IC   标准化集成电路
9. semiconductor device fabrication   半导体设备制造
10. small-scale integration   小规模集成
11. medium-scale integration   中规模集成
12. large-scale integration   大规模集成
13. very large-scale integration   超大规模集成
14. ultra-large-scale integration   极大规模集成

Reference Translation 参考译文

1. 定义

集成电路或单片集成电路（也称为 IC 或芯片）是在通常为硅的一小块半导体材料（或"芯片"）上的一组电子电路。图 5-7 为芯片的示意图。

将大量的微型晶体管集成到一个小芯片中会产生比由分立电子元器件构成的电路更小、

更便宜、更快的电路。集成电路的大规模生产能力和可靠性确保了标准的集成电路能够快速取代分立的晶体管电路。集成电路现在几乎应用于所有电子设备中,并使电子世界发生了革命性的变化。

2. 发展

20 世纪中叶,随着半导体器件制造技术的进步,集成电路得以应用。自 20 世纪 60 年代开始,芯片的尺寸、速度和容量都有了很大的进步。这些进步大致遵循摩尔定律,即集成电路中的晶体管数量大约每两年翻一番。

早期的集成电路,每个芯片只有几个晶体管,集成度低意味着设计过程相对简单。按照今天的标准,制造业的收益率也相当低。随着技术的进步,数以百万计,甚至数以亿计的晶体管可以放在一个芯片上,好的设计需要周密的规划。表 5-1 显示了不同代的集成电路。

表 5-1 集成电路的发展历程

简称	名 称	年份	晶体管的数量	逻辑门的数量
SSI	小规模集成	1964	1~10	1~12
MSI	中规模集成	1968	10~500	13~99
LSI	大规模集成	1971	500~20000	100~9999
VLSI	超大规模集成	1980	20000~1000000	10000~99999
ULSI	极大规模集成	1984	>1000000	>100000

3. 优缺点

与分立元器件电路相比,集成电路有两个主要优点:低成本和高性能。

1)成本低是因为芯片及其所有组件都是作为一个单元被印制的,而不是一次只制造一个晶体管。此外,封装的集成电路比分立电路使用的材料更少。

2)由于集成电路元器件体积小,距离近,开关速度快,功耗相对较低,所以性能较高。

集成电路的主要缺点是设计和制造所需光掩模的成本很高。初始成本高意味着只有在预计产量较高的情况下,才应采用集成电路。

Unit 6 Control Theory

控 制 理 论

Part 1 Composition of the Control System
控制系统构成

1. Definition

A control system manages, commands, directs, or regulates the behavior of other devices or systems. It can range from a single home heating controller of a domestic boiler to large industrial control systems.

Figure 6-1 shows an automatic temperature control system. It consists of a sensor to measure the temperature, a controller and a power regulator. The controller compares the measured temperature with the desired temperature, and regulates the output power to make them the same.

Video 20

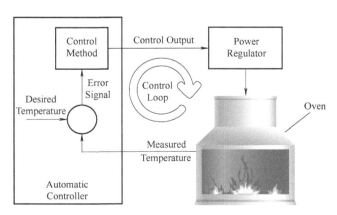

Figure 6-1 An Automatic Temperature Control System

Unit 6 Control Theory

控制理论

2. Classification

Based on different criteria, we can classify the control systems into the following types.

(1) Continuous-time and Discrete-time Control Systems In continuous-time control systems, all the signals are continuous in time. But, in discrete-time control systems, there exists one or more discrete time signals.

(2) Single Input and Single Output(SISO) and Multiple Inputs and Multiple Outputs(MIMO) Control Systems SISO control systems have one input and one output. Whereas, MIMO control systems have more than one input and more than one output.

(3) Open-loop and Closed-loop Control Systems Control Systems can be classified as open-loop control systems and closed-loop control systems based on the feedback path. The details of the open-loop and closed-loop control systems will be described in Part 2.

3. Requirements of a Good Control System

(1) Accuracy Accuracy depends on the measurement tolerance of the system. To increase the accuracy of any control system, an error detector should be present in the control system.

(2) Stability Stability is an important characteristic of the control system. For the bounded input signal, the output must be bounded and if the input is zero then output must be zero.

(3) Speed Speed indicates the time taken by the control system to achieve its stable output. The transient period for a good control system is very small.

Vocabulary 词汇

1. regulate   ['regjuleɪt]   v. 调节，控制
2. sensor   ['sensə(r)]   n. 传感器
3. classification   [ˌklæsɪfɪ'keɪʃn]   n. 分类，类别
4. criteria   [kraɪ'tɪəriə]   n. 标准，准则
5. exist   [ɪg'zɪst]   v. 存在；实际上有
6. accuracy   ['ækjərəsi]   n. 精确度；准确性
7. tolerance   ['tɒlərəns]   n. 公差
8. stability   [stə'bɪləti]   n. 稳定性
9. characteristic   [ˌkærəktə'rɪstɪk]   n. 特点，特征，特性
10. stable   ['steɪbl]   adj. 稳定的，稳固的
11. control   [kən'trəʊl]   v. 指挥；控制
12. automatic   [ˌɔːtə'mætɪk]   adj. 自动的

Notes 注释

1. control system   控制系统
2. open loop control system   开环控制系统
3. closed loop control system 闭环控制系统
4. feedback path   反馈路径
5. transient period   瞬态期

6. power regulator　　　　　　　　　　　　功率调节器
7. output power　　　　　　　　　　　　　输出功率
8. measurement tolerance　　　　　　　　　测量公差
9. continuous-time system　　　　　　　　 连续时间系统
10. discrete-time system　　　　　　　　　离散时间系统
11. Single Input and Single Output (SISO) system　　单输入单输出系统
12. Multiple Inputs and Multiple Outputs (MIMO) system　　多输入多输出系统

Reference Translation　参考译文

1. 定义

控制系统用于管理、命令、指导或调节其他设备或系统的行为。小到一个家用水壶的加热控制器，大到一个大型工业控制系统，都属于控制系统的范畴。

图6-1显示了一个自动温度控制系统。它由测量温度的传感器、控制器和功率调节器组成。控制器将测量温度与所需温度进行比较，并调节输出功率使测量温度和所需温度相同。

2. 分类

根据不同的分类标准，我们可以将控制系统分为以下几种类型。

（1）连续时间和离散时间控制系统　在连续时间控制系统中，所有的信号在时间上都是连续的。但是，在离散时间控制系统中，存在一个或多个离散时间信号。

（2）单输入单输出（SISO）和多输入多输出（MIMO）控制系统　SISO控制系统有一个输入和一个输出，而MIMO控制系统有多个输入和多个输出。

（3）开环控制系统和闭环控制系统　根据是否有反馈路径，控制系统可分为开环控制系统和闭环控制系统。我们将在第2部分中描述开环控制系统和闭环控制系统的细节。

3. 良好控制系统的要求

（1）精度　精度取决于系统的测量公差。为了提高控制系统的精度，控制系统中都应该有误差检测器。

（2）稳定性　稳定性是控制系统的一个重要特性。对于有界输入信号，输出必须有界，如果输入信号为零，则输出信号也必须为零。

（2）响应速度　速度表示控制系统达到稳定输出所需的时间。一个好的控制系统的瞬态周期很小。

Part 2　Open-Loop and Closed-Loop Control
开环和闭环控制

Control Systems can be classified as open-loop control systems and closed-loop(feedback) control systems based on the feedback path.

1. Open-Loop Control

（1）Definition　In open-loop control, the controller that computes its input into a system using

Unit 6 Control Theory
控 制 理 论

only the current state and its model of the system. A characteristic of the open-loop controller is that it does not use feedback to determine if its output has achieved the desired goal of the input.

An example of open-loop controller is a central heating boiler controlled only by a timer. Heat is applied for a constant time, regardless of the temperature of the building.

Video 21

(2) Principle Figure 6-2 shows the block diagram of the open-loop control system. Here, an input is applied to a controller and it produces an actuating signal. This signal is given as an input to a plant which is to be controlled.

Figure 6-2 Open-Loop Control System

2. Closed-Loop Control

(1) Definition A closed loop control system is a set of mechanical or electronic devices that automatically regulates a process variable to a desired state or set point without human interaction.

In the case of the closed-loop control of boiler, a thermostat compares the building temperature with the temperature set on the thermostat. Based on the comparison, the system generates a controller output to maintain the building at the desired temperature by switching the boiler on and off.

A Closed-Loop controller has a feedback loop which manipulates the process variable to be the same as the reference input. For this reason, closed-loop controllers are also called feedback controllers.

(2) Principle Figure 6-3 shows the block diagram of negative feedback Closed-Loop control system. The error detector produces an error signal, which is the difference between the input and the feedback signal. This feedback signal is applied as an input to a controller. The controller produces an actuating signal which controls the plant. In this combination, the output of the control system is adjusted automatically till we get the desired response.

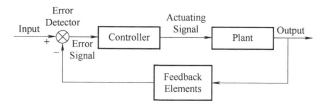

Figure 6-3 Closed-Loop Control System

3. Open-Loop Vs. Closed-Loop Control System

Comparison between the open-loop and closed-loop control systems is shown in Table 6-1.

Table 6-1 Comparison Between Open-Loop and Closed-Loop Control Systems

System	Advantages	Disadvantages
Open-loop Control	1. Simple structure, low cost 2. Easy to regulate	1. Low accuracy and resistance to disturbance 2. Limitations on application
Closed-loop Control	1. Ability to correct error 2. High accuracy and resistance to disturbance	1. Complex structure, high cost 2. Selecting parameter is critical (may cause stability problem)

The primary advantage of a closed-loop feedback control system is its ability to reduce a system's sensitivity to external disturbances, giving the system a more robust control.

The main disadvantage of a closed-loop control system is that in order to provide the required amount of control, a closed-loop system must be more complex than an open-loop system, by having one or more feedback paths.

Vocabulary　词汇

1. determine　[dɪˈtɜːmɪn]　v. 查明，弄清，确定
2. thermostat　[ˈθɜːməstæt]　n. 恒温器
3. combination　[ˌkɒmbɪˈneɪʃn]　n. 结合体，混合体
4. parameter　[pəˈræmɪtə(r)]　n. 参数
5. critical　[ˈkrɪtɪkl]　adj. 关键的，重要的
6. primary　[ˈpraɪməri]　adj. 主要的，最重要的；基本的，最初的
7. sensitivity　[ˌsensəˈtɪvəti]　n. 敏感性，灵敏性
8. external　[ɪkˈstɜːnl]　n. 外部；外部情况 adj. 外部的，外面的，外界的
9. disturbance　[dɪˈstɜːbəns]　n. 干扰；障碍；紊乱
10. feedback　[ˈfiːdbæk]　n. 反馈的意见；（电器的）反馈噪声
11. timer　[ˈtaɪmə(r)]　n. 时计；计时器；定时器
12. manipulate　[məˈnɪpjuleɪt]　v. 控制，操纵，影响

Notes　注释

1. actuating signal　　　　　　　　　　　　　执行信号
2. negative feedback closed-loop control system　负反馈闭环控制系统
3. robust control　　　　　　　　　　　　　鲁棒控制
4. feedback path　　　　　　　　　　　　　反馈通道
5. current state　　　　　　　　　　　　　当前状态
6. process variable　　　　　　　　　　　　过程变量
7. set point　　　　　　　　　　　　　　　设定点
8. feedback loop　　　　　　　　　　　　　反馈回路
9. reference input　　　　　　　　　　　　参考点
10. feedback controller　　　　　　　　　　反馈控制器

11. error detector		误差检测器
12. error signal		误差信号
13. resistance to disturbance		抗干扰性

Reference Translation 参考译文

控制系统可根据是否存在反馈路径分为开环控制系统和闭环控制系统。

1. 开环控制

（1）定义 在开环控制中，控制器只使用系统的当前状态及其模型来计算系统的控制输入。开环控制器的一个特点是，它不使用反馈来确定系统输出是否达到了期望的输入目标。

开环控制器的一个例子是仅由定时器控制的中央供暖锅炉。供暖锅炉定时加热，不管建筑物的温度如何。

（2）原理 图6-2显示了开环控制系统的框图。图中，信号输入到控制器后产生一个执行信号或控制信号。该信号可以作为输入信号提供给要控制的设备。

2. 闭环控制

（1）定义 闭环控制系统是一套机械或电子装置，它能自动将过程变量调节到所需的状态或设定点，而无需人工干预。

在锅炉闭环控制的例子中，恒温器将建筑物温度和恒温器上设定的温度进行比较，根据比较结果得到误差值，闭环控制系统将生成一个控制器输出，通过打开和关闭锅炉的操作将建筑物维持在所需温度。

闭环控制器具有反馈回路，该回路确保控制器施加控制动作以操纵过程变量与参考输入相同。因此，闭环控制器也称为反馈控制器。

（2）原理 图6-3为负反馈闭环控制系统的框图。图中，误差检测器产生误差信号，即输入信号和反馈信号之间的差。反馈信号作为控制器的输入。控制器产生一个控制设备的驱动信号。在这种组合中，控制系统的输出会被自动调整，直到我们得到所需的响应。

3. 开环控制系统和闭环控制系统对比

开环控制系统和闭环控制系统的对比见表6-1。

表6-1 开环控制系统和闭环控制系统的对比

系统	优 点	缺 点
开环控制	1. 结构简单，成本低 2. 容易调试	1. 低精度，低抗干扰性 2. 应用场合少
闭环控制	1. 具有误差校正能力 2. 高精度，高抗干扰性	1. 结构复杂，成本高 2. 参数选择很关键（不适当的参数会造成系统不稳定）

闭环反馈控制系统的主要优点是能够降低系统对外部干扰的敏感性，使系统具有更强的鲁棒性。

闭环反馈控制系统的主要缺点是，为了提供所需的控制量，闭环系统必须包括一个或多个反馈路径，这使得它的结构变得更加复杂。

PID Control
PID 控制

1. Definition

A proportional-integral-derivative(PID) controller is a control loop feedback mechanism widely used in industrial control systems. A PID controller continuously calculates an error value $e(t)$ as the difference between a desired set point and a measured process variable. Then it applies a correction based on proportional, integral, and derivative terms (denoted P, I, and D respectively), hence the name.

2. Principle

The distinguishing feature of the PID controller is the ability to apply accurate and optimal control by using the three control terms of proportional, integral and derivative.

Figure 6-4 shows a block diagram of a PID controller, where $r(t)$ is the desired process variable or setpoint, and $y(t)$ is the measured process variable. An error value $e(t)$ is computed as the difference between a desired setpoint $r(t)$ and a measured process variable $y(t)$. The controller attempts to minimize the error by adjusting a control variable $u(t)$ to a new value determined by the weighted sum of the control terms.

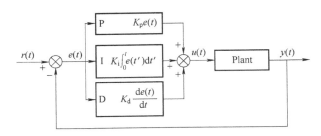

Video 22

Figure 6-4　A Block Diagram of a PID Controller in a Feedback Loop.

In this model:

1) Term P is proportional to the current value of the error $e(t)$. For example, if the error is large and positive, the control output will be proportionately large and positive.

2) Term I accounts for past values of the error $e(t)$ and integrates them over time. For example, if there is a residual error $e(t)$ after the application of proportional control, the integral term seeks to eliminate the residual error by adding a control effect due to the historic cumulative error.

3) Term D is a best estimate of the future trend of the error $e(t)$, based on its current rate of change. It is sometimes called "anticipatory control", as it is effectively seeking to reduce the effect of the error $e(t)$ by exerting a control influence generated by the changing rate of the error. The more rapid the change, the greater the controlling effect.

3. Mathematical Form

The overall control function can be expressed mathematically as

Unit 6 Control Theory
控制理论

$$u(t) = K_p e(t) + K_i \int_0^t e(t')\,dt' + K_d \frac{de(t)}{dt}$$

where K_p, K_i, and K_d, all non-negative, denote the coefficients for the proportional, integral, and derivative terms respectively (sometimes denoted P, I, and D).

Vocabulary　词汇

1. proportional　　　[prə'pɔːʃnl]　　　adj. 相称的；成比例的
2. integral　　　[ɪn'tegrəl]　　　n. 积分
3. derivative　　　[dɪ'rɪvətɪv]　　　n. 微分
4. calculate　　　['kælkjʊleɪt]　　　v. 计算，核算；推测
5. respectively　　　[rɪ'spektɪvli]　　　adv. 分别地；各自地
6. distinguishing　　　[dɪ'stɪŋgwɪʃɪŋ]　　　adj. 有区别的
7. residual　　　[rɪ'zɪdjuəl]　　　adj. 剩余的；残留的
8. eliminate　　　[ɪ'lɪmɪneɪt]　　　v. 排除，清除；淘汰
9. anticipatory　　　[æn,tɪsɪ'peɪtəri]　　　adj. 期望的；预期的
10. effectively　　　[ɪ'fektɪvli]　　　adv. 有效地
11. mathematical　　　[,mæθə'mætɪkl]　　　adj. 数学的
12. coefficient　　　[,kəʊɪ'fɪʃnt]　　　n. 系数

Notes　注释

1. proportional-integral-derivative (PID) controlle　　　比例积分微分控制器
2. feedback mechanism　　　反馈机制
3. non-negative　　　非负的
4. optimal control　　　最优控制
5. control variable　　　控制变量
6. residual error　　　残差
7. cumulative error　　　累计误差
8. changing rate of the error　　　误差变化率

Reference Translation　参考译文

1. 定义

比例-积分-微分控制器（PID 控制器）是一种控制回路反馈机制，广泛应用于工业控制系统。PID 控制器连续计算设定点和测量过程变量之间的差异，作为误差值 $e(t)$，并使用比例、积分和微分环节（分别表示 P、I 和 D）对误差进行校正，因此得名。

2. 原理

PID 控制器的显著特征是能够利用比例、积分和微分三个控制环节影响控制器的输出，从而实现精确的最优控制。图 6-4 显示了 PID 控制器的框图，其中 $r(t)$ 是所需的过程值或设定点，$y(t)$ 是测量的过程值。误差值 $e(t)$ 是设定点 $r(t)$ 和测量过程变量 $y(t)$ 之间的

差值。控制器试图通过将控制变量 $u(t)$ 调整到控制项的加权和确定的新值来减小误差。

在此模型中:

1) P 比例环节与误差 $e(t)$ 的当前值成正比例。例如,如果误差较大且为正值,则控制输出将按比例地放大且为正值。

2) I 积分环节考虑了误差 $e(t)$ 的过去值,并对其取时间的积分。例如,如果在应用比例控制后存在残余误差 $e(t)$,则积分环节通过误差的历史累积值产生的控制效果来消除残余误差。

3) D 微分环节是基于当前变化率对误差 $e(t)$ 未来趋势的最佳估计。它有时被称为"预期控制",因为它通过误差变化率所产生的影响进行控制,来寻求有效地降低误差 $e(t)$ 的影响。这种变化越快,D 微分环节的控制或抑制效果就越明显。

3. 数学形式

PID 控制器的整体控制功能可以用数学表示为

$$u(t) = K_p e(t) + K_i \int_0^t e(t') \, \mathrm{d}t' + K_d \frac{\mathrm{d}e(t)}{\mathrm{d}t}$$

其中 K_p,K_i 和 K_d 均为非负,分别表示比例项、积分项和微分项的系数(有时表示 P、I 和 D)。

Intelligent Control
智能控制

Intelligent control is a class of control techniques that use various artificial intelligence computing approaches like machine learning, artificial neural networks, fuzzy control and expert system.

Video 23

1. Machine Learning

Machine Learning(ML) is an interdisciplinary subject involving statistics, system identification, approximation theory, neural network, optimization theory, computer science, brain science and many other fields. It has a deep connection with pattern recognition, statistical learning, data mining, computer vision, speech recognition, natural language processing and other fields, as shown in Figure 6-5.

Machine learning control is a subfield of machine learning, intelligent control and control theory. It solves optimal control problems with methods of machine learning. Key applications are complex nonlinear systems for which linear control theory methods are not applicable.

Machine learning control has been successfully applied to many nonlinear control problems, for example, attitude control of satellites, building thermal control, and remotely operated under water vehicle etc.

2. Artificial Neural Networks

Artificial neural networks(ANN) are computing systems inspired by the biological neural networks that constitute animal brains. The neural network itself is not an algorithm, but rather a frame-

Unit 6　Control Theory

控 制 理 论

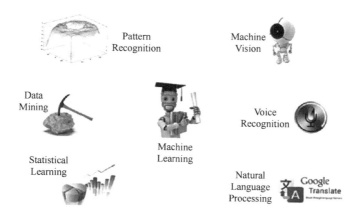

Figure 6-5　Machine Learning and Related Disciplines

work for many different machine learning algorithms to work together. Such systems "learn" to perform tasks by considering examples, generally without being programmed with any task-specific rules.

Artificial neural networks have been used on a variety of tasks, including computer vision, speech recognition, machine translation, social network filtering, playing board and video games and medical diagnosis.

3. Fuzzy Control

Fuzzy control is a theoretical control method based on the basic idea of fuzzy mathematics. In the traditional control field, the accuracy of the dynamic mode of control system is the most important factor affecting the quality of control. The more detailed the dynamic information of the system, the more precise the control can be achieved. However, for complex systems, it is difficult to accurately describe the dynamics of the system, due to too many variables. In other words, the traditional control theory has strong and powerful control ability for definite systems, but it is powerless for systems that are too complex or difficult to describe accurately.

Therefore, people try to deal with these control problems with fuzzy mathematics. Figure 6-6 shows an example of fuzzy logic control of air conditioner. The fuzzy logic system inside the air conditioner reads both the room temperature and the target temperature. Then based on the fuzzy logic, the system will give proper heat/cool/no change command to the air conditioner.

4. Expert System

In artificial intelligence, an expert system is a computer system that emulates the decision-making ability of a human expert. Expert systems are designed to solve complex problems by reasoning through bodies of knowledge, represented mainly as if-then rules rather than through conventional procedural code.

The first expert systems were created in the 1970s and then proliferated in the 1980s. Expert systems were among the first truly successful forms of artificial intelligence software.

An expert system is divided into two subsystems: the inference engine and the knowledge base. The knowledge base represents facts and rules. The inference engine applies the rules to the known facts to deduce new facts. Inference engines can also include explanation and debugging abilities. An expert system is shown in Figure 6-7.

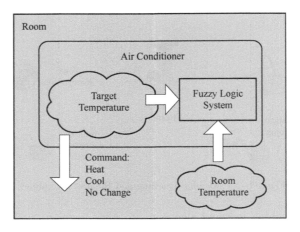

Figure 6-6 Fuzzy Logic Control of Air Conditioner

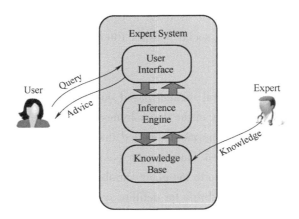

Figure 6-7 An Expert System

Vocabulary 词汇

1. intelligent	[ɪnˈtelɪdʒənt]	adj.	聪明的；智能的
2. technique	[tekˈniːk]	n.	技巧；工艺；技术
3. approach	[əˈprəʊtʃ]	n. 方法，方式，途径 v. 靠近，接近	
4. neural	[ˈnjʊərəl]	adj.	神经的，神经系统的
5. statistic	[stəˈtɪstɪk]	n.	统计学
6. subfield	[sʌbfiːld]	n.	分区；分支；分支学科
7. method	[ˈmeθəd]	n.	方法，办法
8. application	[æplɪˈkeɪʃn]	n.	申请；应用，运用
9. nonlinear	[nɒnˈlɪnɪə]	adj.	非线性的

10. remotely	[rɪˈməʊtli]	adv.	远程地；遥遥地
11. inspire	[ɪnˈspaɪə(r)]	v.	激励，鼓舞；启发
12. biological	[ˌbaɪəˈlɒdʒɪkl]	adj.	生物的；生物学的
13. constitute	[ˈkɒnstɪtjuːt]	v.	组成，构成
14. framework	[ˈfreɪmwɜːk]	n.	框架，结构
15. variable	[ˈveəriəbl]	adj. 可变的，多变的 n. 变量，可变因素
16. definite	[ˈdefɪnət]	adj. 肯定的，确定的 n. 肯定的事（或人）
17. target	[ˈtɑːgɪt]	n. 目标，对象 v. 面向；把……作为攻击目标
18. emulate	[ˈemjuleɪt]	v.	仿真，模仿
19. procedural	[prəˈsiːdʒərəl]	adj.	程序上的；程序性的
20. proliferate	[prəˈlɪfəreɪt]	v.	迅速繁殖（或增值）；猛增
21. subsystem	[sʌbˈsɪstəm]	n.	子系统；子体系；子系统模块
22. debug	[ˌdiːˈbʌg]	v.	排错，调试

Notes 注释

1. fuzzy control	模糊控制
2. pattern recognition	模式识别
3. voice recognition	声音识别
4. statistical learning	统计学习
5. expert system	专家系统
6. system identification	系统识别
7. optimization theory	优化理论
8. data mining	数据挖掘
9. attitude control	姿态控制
10. artificial neural network	人工神经网络
11. a variety of	种种，各种各样的
12. medical diagnosis	医疗诊断
13. artificial intelligence	人工智能
14. inference engine	推理机

Reference Translation 参考译文

　　智能控制是一类使用各种人工智能计算方法的控制技术，如机器学习、人工神经网络、模糊控制和专家系统。

1. 机器学习

　　机器学习（ML）是一门涉及统计学、系统辨识、近似理论、神经网络、优化理论、计算机科学、脑科学等诸多领域的交叉学科。它跟模式识别、统计学习、数据挖掘、计算机视觉、语音识别和自然语言处理等领域有着很深的联系，如图6-5所示。

　　机器学习控制是机器学习、智能控制和控制理论的一个分支。它用机器学习的方法来解决最优控制问题。机器学习控制的关键应用领域是线性控制理论方法不适用的复杂的非线性系统。

机器学习控制已经成功地应用于许多非线性控制问题，比如卫星姿态控制、建筑物温度控制和遥控水下车辆等。

2. 人工神经网络

人工神经网络（ANN）是受构成动物大脑的生物神经网络启发的计算系统。神经网络本身不是一种算法，而是许多不同机器学习算法协同工作的框架。此类系统通过示例来"学习"执行任务，通常不使用任何特定于任务的规则进行编程。

人工神经网络已广泛应用于计算机视觉、语音识别、机器翻译、社交网络过滤、游戏机和电子游戏以及医学诊断等领域。

3. 模糊控制

模糊控制是基于模糊数学基本思想的一种理论控制方法。在传统的控制领域，控制系统动态模式的精度是影响控制质量的最主要关键。系统动态信息越详细，则越能达到精确控制的目的。然而，对于复杂的系统，由于变量太多，往往难以正确地描述系统的动态。换言之，传统的控制理论对于动态信息明确的系统有强有力的控制能力，但对过于复杂或难以精确描述的系统，则显得无能为力了。

因此，人们便尝试着以模糊数学来处理这些控制问题。图6-6所示为空调模糊逻辑控制实例。空调内部的模糊逻辑系统同时读取室温和目标温度，然后，根据模糊逻辑，系统将向空调发出适当的加热、制冷、无变化等指令。

4. 专家系统

在人工智能中，专家系统是模拟人类专家决策能力的计算机系统。专家系统的设计目的是通过知识体系的推理来解决复杂的问题，这些知识主要表现为 if-then 规则，而不是传统的程序代码。

第一批专家系统创建于20世纪70年代，然后在80年代专家系统的数量激增。专家系统是第一批真正成功的人工智能软件之一。

专家系统分为推理机和知识库两个子系统。知识库代表事实和规则。推理机将规则应用于已知的事实以推断新的事实。推理机还可以具有解释和调试能力。

Unit 7 Sensors and Measurements Technology

传感检测技术

Part 1 Resistive Sensor and Capacitive Sensor
电阻式传感器和电容式传感器

In the broadest definition, a sensor is a device, or module whose purpose is to detect events or changes in its environment. A sensor is always used with other electronics.

1. Definition of Resistive Sensor

Resistive sensors are sensors whose resistance vary because of the environmental changes. The change in resistance is measured by electrical measuring devices. Resistive sensors are used for measuring the physical quantities like temperature, displacement, vibration etc.

2. Principle of Resistive Sensor

The resistance of a resistor is given by:

$$R = \rho L/A$$

Video 24

where R is the resistance in ohms, A is the cross-section area of the conductor in meter squares, L is the length of the conductor in meters, and ρ is the resistivity of the conductor in materials in ohm meter. The resistance of a resistive sensor changes if the length, area or resistivity of the metal changes.

3. Types of Resistive Sensor

The following are two types of the resistive sensor.

(1) **Sliding contact devices** One of the most popular sliding contact variable resistance sensors is potentiometer. Schematic symbols of a potentiometer are shown in Figure 7-1.

(2) **Wire resistance strain gauge** This is a device used for the measurement of force, stress

and strain. One example of wire resistance strain gauge is shown in Figure 7-2.

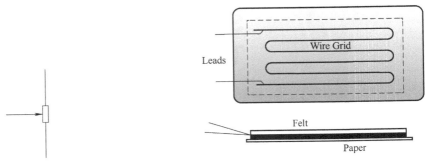

Figure 7-1 Schematic Symbols of a Potentiometer   Figure 7-2 Wire Resistance Strain Gauge

4. Definition of Capacitive Sensor

The capacitive sensor is used for measuring the displacement, pressure and other physical quantities. It is a passive sensor which requires external power for operation. The capacitive sensor works on the principle of variable capacitances.

5. Principle of Capacitive Sensor

A capacitor consists of two conductors separated by a non-conductive region. A voltage U between the two conductors causes electric charges and an electric field. The capacitance C is defined by $C = Q/U$.

The capacitance changes if the distance between the plates or the overlap of the plates changes, as shown in Figure 7-3.

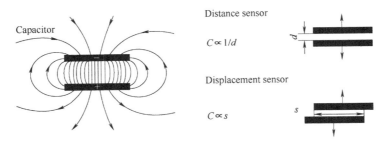

Figure 7-3 Principle of Capacitive Sensor

If the input displacement d decreases, the capacity C of a distance sensor increases.

If the overlap s of the capacity plates increases, the capacity C of a distance sensor increases.

6. Application of Capacitive Sensor

The following are some applications of the capacitive sensor.

1) Measurement of both the linear and angular displacement.

2) Measurement of the force and pressures.

Vocabulary 词汇

1. displacement   [dɪs'pleɪsmənt]   n. 取代；位移

Unit 7 Sensors and Measurements Technology

传感检测技术

2. vibration　　　　　[vaɪˈbreɪʃn]　　　　n. 振动，抖动
3. resistivity　　　　[ˌriːzɪˈstɪvəti]　　　n. 电阻系数；抵抗力；电阻率
4. potentiometer　　[pəˌtenʃiˈɒmɪtə(r)]　n. 电位器；电势差计；电位计
5. schematic　　　　[skiːˈmætɪk]　　　　adj. 略图的；图解的；概要的
　　　　　　　　　　　　　　　　　　　　n. 简图；原理图
6. separate　　　　　[ˈsepəreɪt]　　　　adj. 独立的；分开的 v. （使）分开；隔开
7. plate　　　　　　[pleɪt]　　　　　　n. 极板
8. resistance　　　　[rɪˈzɪstəns]　　　　n. 电阻

Notes 注释

1. resistive sensor　　　　　　　　电阻式传感器
2. capacitive sensor　　　　　　　 电容式传感器
3. physical quantity　　　　　　　 物理量
4. sliding contact device　　　　　 滑动接触装置
5. wire resistance strain gauge　　 线电阻应变片
6. non-conductive　　　　　　　　非导电的
7. angular displacement　　　　　 角位移
8. electrical measuring device　　 电气测量装置
9. variable capacitance　　　　　　可变电容

Reference Translation 参考译文

从广义上来讲，传感器是一种能够检测环境中的事件或变化的设备或模块。传感器通常与其他电子设备一起配合使用。

1. 电阻式传感器的定义

传感器的电阻值因环境变化而变化。电阻式传感器的电阻变化能够被电气装置所检测。电阻式传感器被用于测量温度、位移、振动等物理量。

2. 电阻式传感器的原理

电阻器的阻值计算公式为

$$R = \rho L / A$$

式中 R 为电阻，以 Ω 为单位；A 为导体的横截面积，以 m^2 为单位；L 为导体长度，以 m 为单位；ρ 为材料中导体的电阻率，以 $\Omega \cdot m$ 为单位。电阻式传感器的电阻随着金属的长度、面积和电阻率的改变而改变。

3. 电阻式传感器的类型

下面列举了两种类型的电阻式传感器。

（1）滑动接触装置　目前最流行的滑动接触式可变电阻传感器之一是电位计。电位计的电路符号如图 7-1 所示。

（2）线电阻应变片　这是一个用于测量力、应力和应变的装置。线电阻应变片的示意图如图 7-2 所示。

4. 电容式传感器的定义

电容式传感器是一种用于测量位移、压力和其他物理量的传感器。电容式传感器是一种无源传感器，需要连接外部电源才能工作。电容式传感器是基于可变电容器的原理来工作的。

5. 电容式传感器的原理

电容器由两个被非导电区域隔开的导体组成。两个导体之间的电压 U 引起电荷 Q 和电场。电容 C 的定义为 $C = Q/V$。

如果极板之间的距离或极板的重叠面积改变，电容就会改变，如图7-3所示。

如果极板间的距离 d 减小，电容式传感器的电容量 C 增加。

如果极板的重叠面积 s 增加，电容式传感器的电容量 C 增加。

6. 电容式传感器的应用

以下是电容式传感器的一些应用。

1）线性位移和角位移的测量。

2）力和压力的测量。

Part 2 Inductive Sensor 电感式传感器

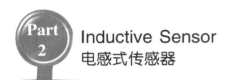

1. Definition

An inductive sensor uses the principle of electromagnetic induction to detect or measure objects. Inductive sensor converts the change of displacement, vibration, pressure, flow rate, rotational speed and metal material into the change of self-inductance or mutual inductance coefficient of equivalent inductance, and then converts the change of inductance into the change of voltage, current and frequency through signal conditioning circuit.

Video 25

Inductive sensor has the advantages of simple structure, reliable operation, long life, high sensitivity and resolution, high accuracy and good linearity. Its main disadvantage is that its frequency response is low and it is not suitable for fast dynamic measurement.

There are many kinds of inductors. This section mainly introduces three types of inductance sensors: self-inductance type(variable reluctance type), mutual inductance type and eddy current type.

2. Classfication

（1）Self-inductance inductive sensor The basic structure of self-inductance inductive sensor is shown in Figure 7-4. It consists of three parts: coil, core and armature. There is an air gap between the core and the armature. The air gap thickness is expressed by the symbol δ. The motion part of the sensor is connected with the armature. During the measurement, the measured armature moves and the thickness of air gap δ changes, which results in the change of magnetoresistance in the magnetic circuit and the change of inductance value L of the inductance coil. Therefore, as long as the change of self-inductance can be measured, the measured value can be determined.

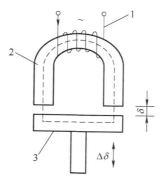

Figure 7-4 Basic Structure of Self-Inductance Inductive Sensor
1—Coil 2—Core 3—Armature

When the permeability of core and derivative is neglected, the inductance value L of coil can be calculated as

$$L = \frac{N^2}{R_m} = \frac{N^2 \mu_0 A_0}{2\delta}$$

where, N——The number of turns of the coil;

R_m——Total Magnetoresistance of air gap;

μ_0——Air gap permeability;

δ——Thickness of air gap;

A_0——Cross section area of air gap.

(2) Mutual-inductance inductive sensor Mutual-inductance inductive sensor is a kind of electromagnetic mechanism which converts non-electric quantity into mutual inductance between coils by using mutual inductance phenomenon in electromagnetic induction. Because it works on the principle of transformer and often uses two secondary coils to form differential type, it is also called differential transformer inductive sensor.

There are many types of differential transformer, such as variable air gap type, variable cross-section area type and solenoid type. At present, solenoid type is widely used in practical work.

Figure 7-5 is the basic structure of the commonly used three-stage solenoid differential transformer. It consists of a primary coil p and two secondary coils s_1 and s_2. The structure size and electrical parameters of the two secondary coils are identical, and the reverse polarity is connected in series. A cylindrical iron core b is inserted into the center of the coil.

(3) Eddy current inductive sensor When high frequency alternating current is applied to inductance coil, alternating current will produce an alternating magnetic field. When metal conductor is placed in the alternating magnetic field, the surface and interior of the conductor will generate induced current. The streamline of the induced current is automatically closed, so it is usually called eddy current. Eddy current inductive sensor is based on eddy current effect. It is also called eddy

current inductance sensor. Its basic structure is shown in Figure 7-6.

Eddy current sensor has been widely used because of its simple structure, easy non-contact continuous measurement, high sensitivity and strong applicability.

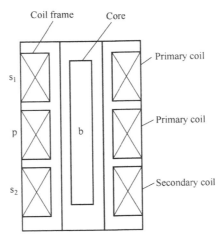

Figure 7-5 Basic Structure of the Commonly Used three-stage solenoid differential transformer

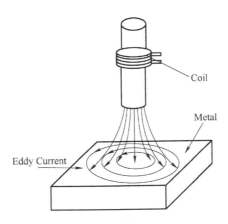

Figure 7-6 Basic Structure of Eddy Current Inductive Sensor

Vocabulary 词汇

1. inductive   [ɪn'dʌktɪv]   adj. 电感应的
2. magnetoresistance   [mæg,niːtəʊviˈzɪstəns]   n. 磁（致电、控电）阻；磁阻效应
3. solenoid   ['sɒlənɔɪd]   n. 螺线管（通电时产生磁场）
4. streamline   ['striːmlaɪn]   v. 使成流线形

Notes 注释

1. inductive sensor   电感式传感器
2. flow rate   流速

3. rotational speed　　　　　　　　　　旋转速度；角速度
4. metal material　　　　　　　　　　　金属材料
5. self-inductance　　　　　　　　　　 自感
6. mutual inductance　　　　　　　　　 互感
7. signal conditioning circuit　　　　 信号调节电路
8. electromagnetic induction　　　　　 电磁感应
9. primary coil　　　　　　　　　　　 一次线圈
10. secondary coil　　　　　　　　　　 二次线圈
11. cross-section area　　　　　　　　 横截面积
12. eddy current　　　　　　　　　　　 涡流
13. magnetic circuit　　　　　　　　　 磁路
14. inductance coil　　　　　　　　　　电感线圈
15. differential transformer inductive sensor　　差动变压器电感式传感器
16. cylindrical iron core　　　　　　　圆柱铁心
17. alternating magnetic field　　　　 交变磁场
18. induced current　　　　　　　　　　感应电流

Reference Translation　参考译文

1. 定义

电感式传感器是使用电磁感应原理来检测或测量对象的设备。电感式传感器将位移、振动、压力、流量、转速、金属材质等被测非电量的变化转换为等效电感的自感或互感系数的变化，再通过信号调整电路将电感的变化转换为电压、电流、频率等电量的变化。

电感式传感器具有结构简单、工作可靠、寿命长、灵敏度高、分辨率高、精度高和线性度好等优点，其主要缺点在于频率响应低，不适用于进行快速动态测量。

电感器的种类也很多，本节主要介绍自感式（变磁阻式）、差动变压器式（互感式）和涡流式这3类电感式传感器。

2. 分类

（1）自感式传感器　自感式传感器的基本结构如图 7-4 所示。它由线圈、铁心和衔铁三部分组成。在铁心与衔铁之间有气隙，气隙厚度用符号 δ 表示，传感器的运动部分与衔铁相连。在进行测量时，被测量带动衔铁移动，使气隙厚度 δ 发生变化，从而引起磁路中的磁阻发生变化，导致电感线圈的自感值 L 变化，因此只要能测出这种自感量的变化，就能确定被测量的大小。

当忽略铁心和衔铁的磁导率时，线圈的自感值 L 可计算为

$$L = \frac{N^2}{R_m} = \frac{N^2 H_0 A_0}{2\delta}$$

式中　N——线圈的匝数；

　　　R_m——空气隙的总磁阻；

　　　μ_0——空气的磁导率；

　　　δ——空气气隙的厚度；

A_0——气隙截面积。

（2）差动变压器式传感器　互感式传感器是利用电磁感应中的互感现象，将非电量转换为线圈间互感的一种电磁机构。由于其是基于变压器原理工作的，而且经常采用两个二次线圈组成差动式，故又称为差动变压器式传感器。

差动变压器式传感器的结构形式较多，有变气隙型、变截面积型和螺管型等，目前在实际工作中，使用得较多的为螺管型。

图 7-5 所示为常用的三段式螺管型差动变压器式传感器的基本结构。它由一个一次线圈 p 和两个二次线圈 s_1 和 s_2 组成，两个二次线圈的结构尺寸和电气参数完全相同，而且反极性串联相接。线圈中心插入圆柱形铁心 b。

（3）涡流式传感器　给电感线圈通以高频交流励磁电流时，交变的电流将会产生一个交变的磁场，当金属导体置于该交变磁场中时，导体表面和内部都会产生感应电流，这种感应电流的流线是自动闭合的，故通常称为电涡流。涡流式传感器就是基于涡流效应而工作的，因此称之为涡流式传感器，其基本结构如图 7-6 所示。

涡流式传感器的结构简单，易于进行非接触的连续测量，而且灵敏度较高，适用性强，因此得到了广泛的应用。

Part 3　Temperature Sensor
温度传感器

1. Definition

The temperature sensor is an electrical device used for automatic measuring of temperature. It converts the thermal energy into a physical quantity likes the displacement, pressure and electrical signal etc.

Video 26

2. Types of Temperature Sensor

The temperature sensor is mainly classified into two types.

（1）Contact temperature sensor device　These types of temperature sensor are required to be in physical contact with the object being sensed and use conduction to monitor changes in temperature. They can be used to detect solids, liquids or gases over a wide range of temperatures.

（2）Non-contact temperature sensor device　The sensing element is not directly contacting the thermal source. They use convection phenomenon for the transfer of heat.

Figure 7-7 shows different types of temperature sensors.

① Thermistor

The Thermistor is a type of temperature sensor, whose name is a combination of the words THERM-ally sensitive res-ISTOR. A thermistor is a special type of resistor which changes its physical resistance when exposed to changes in temperature.

Thermistors are generally made from ceramic materials such as oxides of nickel, manganese or

Unit 7 Sensors and Measurements Technology

传感检测技术

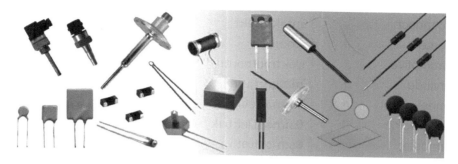

Figure 7-7 Examples of Temperature Sensors

cobalt coated in glass which makes them easily damaged. Their main advantage over snap-action types is their speed of response to any changes in temperature, accuracy and repeatability.

② Thermostat

The Thermostat is a contact type electro-mechanical temperature sensor or switch, that basically consists of two different metals such as nickel, copper, tungsten or aluminum etc., that are bonded together to form a Bi-metallic strip. The different linear expansion rates of the two dissimilar metals produces a mechanical bending movement when the strip is subjected to heat.

③ Thermocouples

Thermocouples consist of a circuit consisting of two different conductors or semiconductors A and B. The two ends of A and B are interconnected, as shown in Figure 7-8.

The measuring end is immersed in the environment whose temperature T_2 has to be measured, while the reference end is held at a different temperature T_1.

An electromotive force will be generated in the circuit. The direction and magnitude of the electromotive force are related to the material of the conductor and the temperature at both ends. This phenomenon is called thermoelectric effect. These two conductors are called thermoelements. The electromotive force generated is called thermoelectric electromotive force.

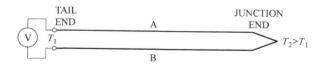

Figure 7-8 Schematic Drawing of a Thermocouple

Because of the temperature difference between junction end and tail end a voltage difference can be measured between the two thermoelements at the tail end: the thermocouple is a temperature-voltage transducer.

Vocabulary 词汇

1. convert	[kən'vɜːt]	v. 转变，转换，转化
2. thermistor	[θɜːˈmɪstə]	n. 热电阻

3. thermocouple [θɜːməʊkʌpl] n. 热电偶
4. interconnect [ˌɪntəkəˈnekt] v. 相互联系，相互连接
5. immerse [ɪˈmɜːs] v. 使沉浸在，置于
6. electromotive [ɪˌlektrəʊˈməʊtɪv] adj. 电动的
7. magnitude [ˈmæɡnɪtjuːd] n. 巨大；重大；重要性
8. thermoelement [ˈθɜːməʊˈelɪmənt] n. 热电极
9. thermoelectric [ˌθɜːməʊɪˈlektrɪk] n. 热电的
10. thermostat [ˈθɜːməstæt] n. 恒温器
11. conduction [kənˈdʌkʃn] n. 热传导
12. oxides [ˈɒksaɪdz] n. 氧化物
13. nickel [ˈnɪkl] n. 镍
14. manganese [ˈmæŋɡəniːz] n. 锰
15. cobalt [ˈkəʊbɔːlt] n. 钴
16. transducer [trænzˈdjuːsə(r)] n. 换能器；变换器

Notes 注释

1. temperature sensor 温度传感器
2. thermal energy 热能
3. contact temperature sensor device 接触式温度传感器
4. non-contact temperature sensor device 非接触式温度传感器
5. convection phenomenon 对流现象
6. ceramic material 陶瓷材料
7. thermally sensitive resistor 热敏电阻
8. electromotive force 电动势
9. thermoelectric effect 热电效应

Reference Translation 参考译文

1. 定义

温度传感器是一种用于自动测量温度的电气设备。它能将热能转换为位移、压力和电信号等物理量。

2. 分类

温度传感器主要分为两种类型。

（1）接触式温度传感器装置 这种类型的温度传感器需要与被感测对象进行物理接触，并使用热传导来监测温度变化。它们可用于在各种温度范围内检测固体、液体或气体的温度。

（2）非接触式温度传感器装置 传感元件不直接接触热源。它们利用对流现象来传递热量。

图 7-7 显示了不同类型的温度传感器。

① 热敏电阻

热敏电阻是一种温度传感器，其名称是"热敏"和"电阻"两个单词的组合。热敏电

阻是一种特殊类型的电阻器，当温度变化时，它的阻值会发生变化。

热敏电阻通常由陶瓷材料制成，如涂有玻璃的镍、锰或钴氧化物，这使得它们很容易被损坏。与快速动作类型的温度传感器相比，它们的主要优势在于对温度变化的快速响应、精确度和可重复性。

② 恒温器

恒温器是一种接触式机电温度传感器或开关，它基本上由两种不同的金属组成，如镍、铜、钨或铝等，它们结合在一起形成一个双金属带。当双金属带受热时，两种不同金属的不同线膨胀系数会产生机械弯曲运动。

③ 热电偶

热电偶包括了由两种不同的导体或半导体 A 和 B 组成的一个回路，其两端相互连接，如图 7-8 所示。把热电偶的两端置于不同的温度。

热电偶一端置于温度 T_2 中，称为测量端（或连接端）；另一端置于温度 T_1 中，称为参考端（或尾端）。

热电偶回路中将产生一个电动势，该电动势的方向和大小与导体的材料及两端点的温度有关。这种现象称为热电效应，这两种导体称为热电极，产生的电动势则称为热电动势。

由于热电偶连接端和尾端之间的温差，可以在尾端测量到两个热电偶之间的电压差。因此，热电偶是一个温度-电压传感器。

Part 4 Photoelectric Sensor
光电式传感器

1. Definition

Photoelectric sensors emit a beam of light to detect the presence or absence of items or changes in surface conditions. When the emitted light is interrupted or reflected by the object, the change in light patterns is measured by a receiver and the target object or surface is recognized.

2. Principle and Major Types

The photoelectric sensor is controlled by converting the change of light into the change of electric signal. In general, photoelectric sensors are composed of three parts: transmitter, receiver and detection circuit. A beam of light is emitted from the light emitting element and is received by the light receiving element.

Video 27

（1）Diffuse-Reflective type Both the light emitting and light receiving elements are contained in a single housing. The sensor receives the light reflected from the target. Principle of the reflective model is shown in Figure 7-9.

（2）Thru-Beam type The transmitter and receiver are separated. When the target is between the transmitter and receiver, the light is interrupted. Principle of the thru-beam model is shown in Figure 7-10.

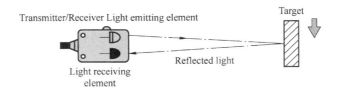

Figure 7-9　Principle of the Diffuse-Reflective Type

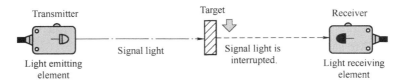

Figure 7-10　Principle of the Thru-Beam Type

(3) Retroreflective type　Both the light emitting and light receiving elements are contained in same housing. The light from the emitting element hits the reflector and returns to the light receiving element. When a target is present, the light is interrupted. Principle of the retroreflective model is shown in Figure 7-11.

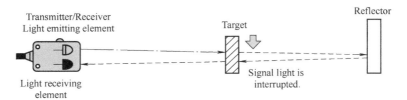

Figure 7-11　Principle of the Retroreflective Type

3. Features

(1) Non-Contact detection　Since detection is possible without contact, there will be no damage to targets. The sensor itself will not be damaged either, ensuring long service life and maintenance-free operation.

(2) Almost all materials detectable　Since the sensor either detects targets based on the reflectivity or the quantity of interrupted light, almost all kinds of materials are detectable. This includes glass, metal, plastic, wood, and liquid.

(3) Long detecting distance　Photoelectric Sensors are generally high power and allow long-range detection.

Vocabulary　词汇

1. interrupt	[ˌɪntəˈrʌpt]	v. 打扰；中断
2. recognize	[ˈrekəgnaɪz]	v. 认识；认出，识别，辨别出；意识到

Unit 7　Sensors and Measurements Technology

传感检测技术

3. transmitter　　　[trænsˈmɪtə(r)]　　　n. 发射器；传送者
4. contain　　　　　[kənˈteɪn]　　　　　v. 包含，含有；控制
5. reflectivity　　　[ˌriːflekˈtɪvɪti]　　　n. 反射率
6. receiver　　　　 [rɪˈsiːvə(r)]　　　　 n. 接收机
7. housing　　　　 [ˈhaʊzɪŋ]　　　　　n.（机器的）外壳，套
8. reflector　　　　[rɪˈflektə(r)]　　　　n. 反光面；反光体

Notes　注释

1. photoelectric sensor　　　　光电式传感器
2. diffuse-reflective type　　　漫射型
3. thru-beam type　　　　　　对射型
4. retroreflective type　　　　 反射型
5. maintenance-free　　　　　不需维护的
6. long-range　　　　　　　　远程地
7. detection circuit　　　　　　检测电路
8. light receiving element　　　光接收元件
9. light emitting element　　　 光发射元件

Reference Translation　参考译文

1. 定义

光电传感器通过发射光束检测物体的存在与否以及表面状况的变化。当发射的光被物体中断或反射时，光的变化被接收器加以检测，从而识别出目标物体或表面。

2. 原理及主要类型

光电传感器是通过将光的变化转化为电信号的变化来工作的。光电传感器一般由发射器、接收器和检测电路三部分组成。光束从发光元件发出，并由光接收元件接收。

（1）漫射型传感器　发光和接收元件都包含在一个外壳中。传感器接收目标反射的光。漫射型传感器的原理如图7-9所示。

（2）对射型传感器　发射器和接收器是分开的。当目标位于发射器和接收器之间时，光束被中断。对射型传感器的原理如图7-10所示。

（3）反射型传感器　发射元件和接收元件都包含在同一个外壳中。来自发射元件的光束撞击反光镜并返回接收元件。当目标出现时，光束被中断。反射型传感器的原理如图7-11所示。

3. 特征

（1）非接触检测　由于非接触检测，因此不会对目标造成损害。传感器本身也不会损坏，确保使用寿命比较长和不需要维护操作。

（2）几乎可检测所有材料　由于传感器根据光的反射和中断来检测目标，因此几乎所有的材料都可以检测到，包括玻璃、金属、塑料、木材和液体。

（3）探测距离长　光电传感器通常都是大功率的，可以进行远程检测。

Part 5 Optical Grating Sensor
光栅传感器

1. Definition and Principle

Optical grating sensor is a kind of sensor which uses optical grating to measure displacement.

The grating is densely spaced parallel lines on a long strip of optical glass with a fringe density of 10-100 lines/mm. The grating fringes have the effect of optical amplification and error averaging. The sensor consists of four parts: ruler grating, indicator grating, optical system and measurement system, as shown in Figure 7-12.

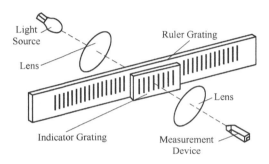

Video 28　　　　Figure 7-12　Components of Optical Grating Transducer

When the ruler grating moves relative to the indicator grating, the overlapping grating fringes of light and shade are formed. The fringes, called Moire fringe, are distributed roughly according to the sinusoidal law, as shown in Figure 7-13.

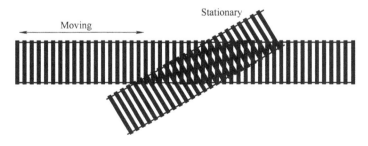

Figure 7-13　Moire Fringe

These fringes move with relative motion of the gratings and directly irradiate the photoelectric elements. A series of electrical pulses are obtained at output terminals of the optical grating sensor. The digital signals are used to measure the displacement.

2. Classification

There are two types of light path of sensor.

(1) Transmission grating　whose grating lines are engraved on transparent materials(such as

Unit 7 Sensors and Measurements Technology

industrial white glass, optical glass, etc.).

(2) Reflection grating whose grating lines are engraved on metal(stainless steel) with strong reflection or glass coated metal(aluminum film).

The advantages of grating sensor are large range and high accuracy. The grating sensor is used in programmable control, numerical control machine tool and three-coordinate measuring mechanism. It can be used to measure static and dynamic linear displacement and circular angular displacement.

Vocabulary 词汇

1. densely	['dɛnsli]	adv.	浓密地，稠密地；密集地
2. fringe	[frɪndʒ]	n.	条纹
3. relative	['relətɪv]	adj.	相对的
4. distribute	[dɪ'strɪbjuːt]	v.	分布；分配；分散
5. roughly	['rʌfli]	adv.	粗糙地；大致，大约
6. irradiate	[ɪ'reɪdɪeɪt]	v.	照射
7. pulse	[pʌls]	n.	脉冲
8. engrave	[ɪn'greɪv]	v.	在……上雕刻
9. transparent	[træns'pærənt]	adj.	透明的；显而易见的
10. coat	[kəʊt]	v.	镀膜
11. aluminum	[ə'luːmɪnəm]	n.	铝

Notes 注释

1. optical glass	光学玻璃
2. measurement system	测量系统
3. according to	据……所说；按照，根据
4. sinusoidal law	正弦定律
5. output terminal	输出终端
6. stainless steel	不锈钢
7. numerical control machine tool	数控机床
8. optical grating sensor	光栅传感器
9. optical amplification	光学放大
10. error averaging	误差平均
11. ruler grating	标尺光栅
12. indicator grating	指示光栅
13. optical system	光学系统
14. measurement system	测量系统
15. grating fringe	光栅条纹
16. Moire fringe	莫尔条纹
17. transmission grating	透射光栅

18. reflection grating　　　　　　　　反射光栅

Reference Translation　参考译文

1. 定义和原理

光栅传感器是一种采用光栅条纹原理测量位移的传感器。

光栅是在一块长条形的光学玻璃上密集等间距分布的刻线，刻线密度为 10～100 线/mm。由光栅形成的叠栅条纹具有光学放大作用和误差平均效应。传感器由标尺光栅、指示光栅、光路系统和测量系统四部分组成，如图 7-12 所示。

当标尺光栅相对于指示光栅移动时，便形成了明暗相间的光栅条纹。这些条纹被称为莫尔条纹，并且大致按正弦规律分布，如图 7-13 所示。

这些条纹随着光栅的相对运动而移动，并直接照射在光电元件上。在光栅传感器的输出端可以得到一系列电脉冲，这些数字信号被用于测量位移。

2. 分类

传感器的光路形式有两种：

（1）透射式光栅　它的光栅线刻在透明材料（如工业用白玻璃、光学玻璃等）上。

（2）反射式光栅　它的光栅线刻在具有强反射的金属（不锈钢）或玻璃镀金属膜（铝膜）上。

光栅传感器的优点是量程大和精度高。光栅传感器可应用在编程控制、数控机床和三坐标测量机构中，可用于测量静态、动态的直线位移和圆角位移。

Part 6　Radio-Frequency identification
射频识别（RFID）

1. Definition

Radio-frequency identification(RFID) uses electromagnetic fields to automatically identify and track tags attached to objects. The tags contain electronically stored information. RFID is one method of automatic identification and data capture.

Video 29

2. Principle

The RFID system is basically composed of electronic tag, reader and data management system. As shown in Figure 7-14. The tag reader is responsible for powering and communicating with a tag. The tag antenna captures energy and transfers the tag's ID(the tag's chip coordinates this process). The encapsulation maintains the tag's integrity and protects the antenna and chip from environmental conditions or reagents.

1）Electronic tag has the function of reading and writing. It exchanges data with reading and writing equipment by radio wave.

2）Reader can transmit the read and write commands of the host computer to the electronic tag, and send the data returned by the electronic tag to the host computer.

3) Data management system stores and manages data information, and reads and writes cards.

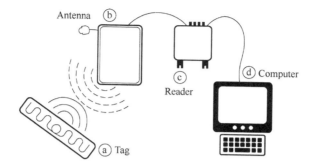

Figure 7-14　Principle of RFID

3. Classification

Many types of RFID exist, but at the highest level, we can divide RFID devices into two classes: active and passive.

Active RFID devices consist active tags, which need a power source—they're either connected to a powered infrastructure or use energy stored in an integrated battery. One example of an active tag is the transponder attached to an aircraft that identifies its national origin.

Passive RFID devices consist passive tags which don't require batteries or maintenance. The tags are small enough to fit into a practical adhesive label.

4. Application

The RFID tag can be affixed to an object and used to track and manage inventory, Figure 7-15 shows a RFID-enabled warehouse for efficiency and productivity.

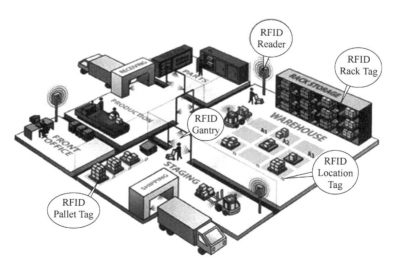

Figure 7-15　RFID-Enabled Warehouse for Efficiency and Productivity

RFID offers advantages over manual systems or use of bar codes. The tag can be read if passed near a reader, even if it is covered by the object or not visible. The tag can be read inside a case, carton, box or other container. Unlike barcodes, RFID tags can be read hundreds at a time. Bar codes can only be read one at a time using current devices.

Vocabulary   词汇

1. automatically   [ˌɔːtəˈmætɪkli]   adv. 自动地；机械地
2. identify   [aɪˈdentɪfaɪ]   v. 确认；识别，辨认出
3. basically   [ˈbeɪsɪkli]   adv. 大体上，基本上
4. infrastructure   [ˈɪnfrəstrʌktʃə(r)]   n. 基础设施
5. battery   [ˈbætri]   n. 电池
6. adhesive   [ədˈhiːsɪv]   n. 黏合剂 adj. 黏合的，黏附的
7. inventory   [ˈɪnvəntri]   n. 库存；清单
8. warehouse   [ˈweəhaʊs]   n. 仓库；货栈
9. efficiency   [ɪˈfɪʃnsi]   n. 效率；效能；功效
10. productivity   [ˌprɒdʌkˈtɪvəti]   n. 生产力；生产率
11. container   [kənˈteɪnə]   n. 集装箱；容器

Notes   注释

1. Radio-frequency identification   射频识别技术
2. data capture   数据捕捉
3. data management system   数据管理系统
4. radio wave   无线电波
5. active tag   有源标签

Reference Translation   参考译文

1. 定义

射频识别技术（RFID）使用电磁场来自动识别和跟踪附着在物体上的标签。标签包含电子存储信息。RFID 是一种自动识别和数据采集的方法。

2. 原理

RFID 系统基本都由电子标签、阅读器和数据管理系统三部分组成，如图 7-14 所示。标签阅读器负责为标签供电并与之通信。标签天线捕获能量并传输标签的 ID（标签的芯片协调此过程）。封装保持标签的完整性，并保护天线和芯片不受环境条件或试剂的影响。

1）电子标签具有读写功能，它通过无线电波与读写设备进行数据交换。

2）阅读器可将主机的读写命令发送给电子标签，并将电子标签返回的数据发送给主机。

3）数据管理系统用于存储及管理数据信息，并对卡进行读写控制等。

3. 分类

目前存在许多类型的 RFID，我们可以将 RFID 设备分为：有源和无源两类。

有源 RFID 设备需要主动标签，主动标签需要电源供电，它们要么连接到一个电源的基

Unit 7 Sensors and Measurements Technology
传感检测技术

础设备上,要么使用集成电池中存储的能量。

无源 RFID 设备包含被动标签,被动标签不需要电池或维护。这些标签足够小,可以被安装在一个可粘贴的标签上。

4. 应用

RFID 标签可以粘贴在物体上,用于跟踪和管理库存。图 7-15 所示为一个使用了 RFID 技术提高效率和生产率的仓库。

RFID 与手动系统或使用条形码相比,具有一定的优势。如果标签在阅读器的附近,即使它被对象覆盖或不可见,也可以被读取。标签可以在纸箱、盒子或其他容器内被读取。与条形码不同的是,一次可以读取数百个 RFID 标签,而条形码只能使用当前设备一次读取一个。

Unit 8 Hydraulics and Pneumatics Transmission

液压与气压传动

Part 1 Hydraulic Transmission System
液压传动系统

1. Definition

A hydraulic transmission system uses pressurized hydraulic fluid to power hydraulic machinery. The term hydraulic transmission refers to the transfer of energy from pressure differences, not from the kinetic energy of the flow.

Video 30

2. Working Principle

Figure 8-1 shows the schematic diagram of the hydraulic system for the working table of a machine tool. Hydraulic pump draws oil from the tank. The pressure oil then flows through the throttle valve, directional control valve to reach the hydraulic cylinder.

When the directional control valve is in the middle position, the working table stops moving; when the handle of the directional control valve changes to the left position, the pressure oil enters the left chamber of the hydraulic cylinder, pushing the piston and driving the working table to move to the right.

When the handle of the directional control valve changes to the right position, the pressure oil enters the right chamber of the hydraulic cylinder, pushing the piston and driving the working table to move to the left.

From the above example, it can be seen that in the hydraulic system, there are two energy transformations: first, the mechanical energy of motor is converted to liquid pressure energy by hydraulic pump, and then the hydraulic pressure energy is converted to mechanical energy by hydraulic cylin-

Unit 8 Hydraulics and Pneumatics Transmission
液压与气压传动

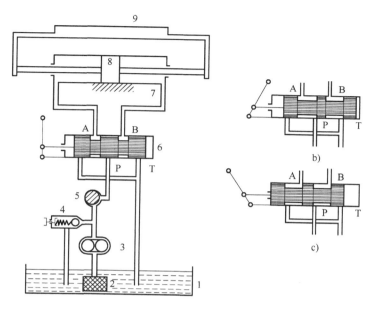

Figure 8-1 Hydraulic Transmission System for the Working Table of a Machine Tool
1—Oil tank 2—Oil filter 3—Oil pump 4—Pressure regulator 5—Throttle valve 6—Directional control valve
7—Hydraulic cylinder 8—Piston 9—Working table

der. The process of hydraulic transmission is the conversion process of mechanical energy-hydraulic energy-mechanical energy.

3. Composition

As can be seen from the above example, the hydraulic transmission system is composed of four parts: power component, actuator component, control component and auxiliary component. The composition of the hydraulic system is shown in Table 8-1.

Table 8-1 Composition of the Hydraulic Transmission System

Components	Devices	Function
Power components	Oil pump	Power components convert the mechanical energy output by the prime mover into liquid pressure energy
Actuator components	Hydraulic cylinder and motor	Actuator converts liquid pressure energy into mechanical energy
Control components	Hydraulic control valves	Control element controls and regulates the pressure, flow rate and flow direction of oil
Auxiliary components	Pipes, tanks, filters, accumulators, etc.	Auxiliary components provide various auxiliary functions including connection, oil transportation, oil storage, filtration, and measurement

4. Symbolic Representation of a Hydraulic System

The schematic diagram shown in Figure 8-1 is intuitive and easy to understand, but is inconvenient to draw, especially when the number of components in the system is large.

Figure 8-2 shows the symbolic representation of the hydraulic transmission system. The symbolic representation is more convenient to draw.

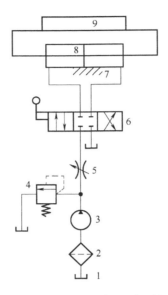

Figure 8-2　Symbolic Representation of the Hydraulic Transmission System
1—Oil tank　2—Oil filter　3—Oil pump　4—Pressure regulator　5—Throttle valve　6—Directional control valve
7—Hydraulic cylinder　8—Piston　9—Working table

Vocabulary　词汇

1. hydraulic	[haɪˈdrɔːlɪk]	adj.	水力的，液压的
2. pressurized	[ˈpreʃəraɪzd]	adj.	增压的；加压的
3. kinetic	[kɪˈnetɪk]	adj.	运动的，活跃的，能动的，有力的
4. schematic	[skiːˈmætɪk]	adj.	略图的；简表的
5. chamber	[ˈtʃeɪmbə(r)]	n.	腔
6. piston	[ˈpɪstən]	n.	活塞
7. regulator	[ˈreɡjuleɪtə(r)]	n.	调整器，校准器，调节器
8. pump	[pʌmp]	n.	泵
9. actuator	[ˈæktjueɪtə]	n.	操作机构；执行机构（元件）；往复运动油（气）
10. symbolic	[sɪmˈbɒlɪk]	adj.	符号的

Notes　注释

1. hydraulic transmission system　　　　　液压传动系统
2. kinetic energy　　　　　　　　　　　　动能
3. pressure oil　　　　　　　　　　　　　压力油

Unit 8　Hydraulics and Pneumatics Transmission
液压与气压传动

4. oil tank	油箱
5. oil filter	滤油器
6. oil pump	油泵
7. directional control valve	换向阀
8. hydraulic cylinder	液压缸
9. pressure regulator	溢流阀
10. throttle valve	节流阀
11. working table	工作台
12. energy transformation	能量转换
13. mechanical energy	机械能
14. liquid pressure energy	液体压力能
15. prime mover	原动机
16. flow rate	流量
17. flow direction	流向

Reference Translation　参考译文

1. 定义

液压传动系统利用加压液压油为液压机械提供动力。术语"液压传动"是指用压力差传递能量，而不是利用液体流动的动能传递能量。

2. 工作原理

图 8-1 所示为机床工作台液压系统的工作原理。液压泵从油箱中吸油，液压油流经节流阀、换向阀，到达液压缸。

当换向阀处于中位时，工作台停止运动；当换向阀的手柄转换到左位时，液压油进入液压缸左腔，推动活塞并带动工作台向右运动。

当换向阀的手柄转换到右位时，液压油进入液压缸右腔，推动活塞并带动工作台向左运动。

从机床工作台的例子可以得到：在液压系统中，要发生两次能量转换，即先通过液压泵把液压马达的机械能转换为液体压力能，再通过液压缸把液体的压力能转换为机械能。液压传动的过程就是机械能—液压能—机械能的转换过程。

3. 组成

由上面的例子还可以看到，液压传动系统都是由动力元件、执行元件、控制元件和辅助元件四部分组成的。液压传动系统的组成见表 8-1。

表 8-1　液压传动系统的组成

组成部分	元　件	作　用
动力元件	液压泵	动力元件将原动机输出的机械能转换成液体压力能
执行元件	液压缸、液压马达	执行元件将液体的压力能转换为机械能
控制元件	液压控制阀	控制元件控制和调节液流的压力、流量和流动方向
辅助元件	油管、油箱、过滤器、蓄能器等	辅助元件起连接、输油、储油、过滤、测量等各种辅助作用

4. 液压传动系统的图形符号

图 8-1 所示的液压传动系统图绘制起来直观、易懂，但是比较麻烦，特别是当系统中元件数量较多时，绘制更加不方便。

图 8-2 所示为用图形符号表示的机床工作台液压传动系统图，这样绘制起来更方便。

Part 2 Features and Application of Hydraulic System
液压系统的特点及应用

1. Advantages and Disadvantages of Hydraulic System

(1) Advantages

1) The hydraulic system uses incompressible fluid which results in high efficiency.

Video 31

2) Since the hydraulic transmission system is connected by tubes, the transmission mechanism can be arranged conveniently and flexibly.

3) It delivers consistent power output which is difficult in mechanical transmission systems.

4) Possibility of leakage is low in hydraulic system. The maintenance cost is low.

5) Hydraulic systems perform well in hot environment conditions.

(2) Disadvantages

1) Because of the large resistance of liquid flow, the efficiency of hydraulic transmission system is low.

2) The material of storage tank, piping, cylinder and piston can be corroded with the hydraulic fluid.

3) The leakage of hydraulic oil will affect the accuracy of the actuator.

4) The structural weight and size of the system is large which makes it unsuitable for the small instruments.

5) The small impurities in the hydraulic fluid can permanently damage the complete system.

2. Application of Hydraulic System

Hydraulic transmission system has been widely used, due to its obvious advantages. The application of hydraulic transmission system in mechanical industry is shown in Table 8-2.

Table 8-2 Application of Hydraulic Transmission System in Mechanical Industry

Industry	Examples
Machine tool industry	Grinding machine, milling machine, broaching machine, press machine
Engineering machinery	Excavators, road roller, bulldozers
Automobiles	Brakes, shock absorbers, steering system
Construction machinery	Pile driver, hydraulic jack, grader

Unit 8　Hydraulics and Pneumatics Transmission
液压与气压传动

(续)

Industry	Examples
Light industry machinery	Packaging machine, injection machine, paper machine
Hoisting and conveying machinery	Truck crane, gantry crane, forklift, belt conveyor

Vocabulary　词汇

1. incompressible　[ˌɪnkəm'presəbl]　adj. 不能（不可，不易）压缩的
2. consistent　[kən'sɪstənt]　adj. 一致的；始终如一的；连续的；持续的
3. mechanical　[mə'kænɪkl]　adj. 机械的，机械学的；手工操作的
4. leakage　['liːkɪdʒ]　n. 泄漏量；渗漏量
5. corrode　[kə'rəʊd]　v. 使腐蚀，侵蚀
6. maintenance　['meɪntənəns]　n. 保养，保管；维护；维修
7. possibility　[ˌpɒsə'bɪləti]　n. 可能；可能性
8. unsuitable　[ʌn'suːtəbl]　adj. 不合适的，不适宜的
9. impurity　[ɪm'pjʊrəti]　n. 杂质；不纯
10. structural　['strʌktʃərəl]　adj. 结构（上）的，构架（上）的
11. permanently　['pɜːmənəntli]　adv. 永久地；永远地
12. excavator　['ekskəveɪtə(r)]　n. 挖掘机
13. bulldozer　['bʊldəʊzə(r)]　n. 推土机
14. brake　[breɪk]　n. 制动器
15. grader　['greɪdə(r)]　n. 平地机
16. forklift　['fɔːklɪft]　n. 叉车

Notes　注释

1. mechanical transmission system　机械传动系统
2. storage tank　储罐
3. machine tool industry　机床制造业
4. grinding machine　磨床
5. milling machine　铣床
6. broaching machine　拉床
7. press machine　冲床
8. road roller　压路机
9. shock absorber　减振器
10. steering system　转向系统
11. construction machinery　建筑机械
12. pile driver　打桩机
13. hydraulic jack　液压千斤顶
14. light industry machinery　轻工业机械

15. packaging machine　　　　　　　　包装机
16. injection machine　　　　　　　　　注塑机
17. paper machine　　　　　　　　　　造纸机
18. hoisting and conveying machinery　　起重运输机械
19. truck crane　　　　　　　　　　　　汽车式起重机
20. gantry crane　　　　　　　　　　　门式起重机
21. belt conveyor　　　　　　　　　　　输送带

Reference Translation　参考译文

1. 液压系统的优缺点

（1）优点

1）液压系统采用不可压缩液体传动，效率高。
2）由于液压传动系统采用油管连接，所以可以方便灵活地布置传动机构。
3）液压系统能够提供稳定的功率输出，这是机械传动系统难以实现的。
4）液压系统发生泄漏的可能性比较小，所以维修费用很低。
5）液压系统在高温环境条件下也能良好运行。

（2）缺点

1）由于液体流动时产生的阻力比较大，所以液压传动系统的效率较低。
2）储液罐、管路、气缸和活塞的材料可能被液压油腐蚀。
3）液压油的泄漏会影响执行元件的准确性。
4）液压系统结构重量和尺寸均较大，不适合小型仪器使用。
5）液压油中的微小杂质会永久损坏整个系统。

2. 液压系统的应用

因为液压传动系统的显著优点，使其得到了普遍的应用。液压传动系统在机械工业中的应用见表8-2。

表8-2　液压传动系统在机械工业中的应用

行业名称	应用场合举例
机床工业	磨床、铣床、冲床、压力机
工程机械	挖掘机、压路机、推土机
汽车工业	制动器、减振器、转向系统
建筑工业	打桩机、液压千斤顶、平地机
轻工业机械	包装机、注塑机、造纸机
起重运输机械	汽车式起重机、门式起重机、叉车、输送带

Unit 8 Hydraulics and Pneumatics Transmission
液压与气压传动

Part 3 Pneumatic Transmission System
气压传动系统

1. Definition

Pneumatic transmission is a technology that uses compressed air as power source to drive and control various mechanical equipment. With the development of industrial mechanization and automation, pneumatic technology is increasingly used in various fields.

2. Working Principle

We will introduce the working principle of pneumatic transmission system through the following example.

Video 32

Figure 8-3 shows the working principle of the pneumatic shearing machine. The diagram shows the situation before shearing. The compressed air generated by the air compressor passes through the air cooler, water separator, air reservoir, air filter, pressure switch, air regulator, and reaches the directional control valve.

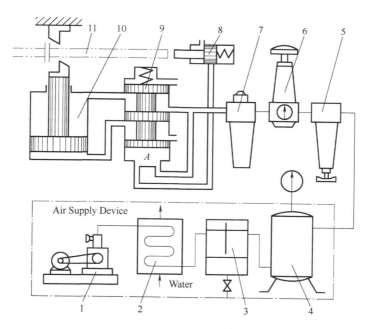

Figure 8-3 Pneumatic Transmission System of Pneumatic Shearing Machine

1—Air compressor 2—Air cooler 3—Water separator 4—Air reservoir 5—Air filter 6—Pressure switch
7—Air regulator 8—Stroke valve 9—Directional control valve 10—Air cylinder 11—Workpiece

At this time, most of the compressed air enters the upper chamber of the cylinder, while the lower chamber of the cylinder connects with the atmosphere, and the piston of the cylinder is in the lowest position.

When the feeding device feeds the workpiece to the specified position, the workpiece presses down the stroke valve, and makes the spool of the directional control valve to move downward. Then compressed air enters the lower chamber of the cylinder, and the upper chamber connects with the atmosphere. The piston of the cylinder moves upward, driving the scissors to cut the workpiece upward.

Figure 8-4 shows the schematic diagram of pneumatic shear pneumatic transmission system drawn with graphic symbols.

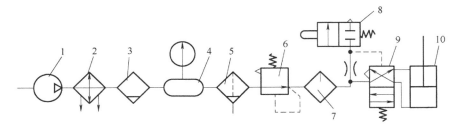

Figure 8-4　Symbolic Representation of Pneumatic Transmission System of Pneumatic Shearing Machine

As can be seen from the example of pneumatic shearing machine, pneumatic transmission system converts the mechanical energy of motor into the pressure energy of gas, and then converts the pressure energy of gas into mechanical energy through cylinder to drive the load to do work. The process of pneumatic transmission is the conversion process of mechanical energy-pneumatic energy-mechanical energy.

3. Composition

As can be seen from the above example, in the pneumatic transmission system, the pneumatic transmission system can be divided into the following four parts according to the functions, as shown in Table 8-3.

Table 8-3　Composition of Pneumatic Transmission System

Components	Devices	Function
Air supply device	Air compressor, air reservoir, air purification device and pipeline, etc.	The air supply device provides compressed air for the system.
Actuator	Cylinder and air motor	The actuator converts the pressure energy of compressed air into the mechanical energy of the working device.
Control components	Pneumatic control valves, such as pressure switch, flow valves, directional control valves and control elements, etc.	Control elements are used to control the pressure, flow rate and flow direction of compressed air.
Auxiliary components	Air regulator, muffler, converter, display, sensor, etc.	Auxiliary components are used for internal lubrication, noise reduction, signal converter, display, amplification, detection, etc.

Unit 8 Hydraulics and Pneumatics Transmission
液压与气压传动

Vocabulary 词汇

1. pneumatic [njuːˈmætɪk] adj. 充气的；气动的
2. mechanization [ˌmekənaɪˈzeɪʃn] n. 机械化；机理
3. atmosphere [ˈætməsfɪə(r)] n. 大气，空气
4. feed [fiːd] n. （机器的）进料
5. workpiece [ˈwɜːkpiːs] n. 工件
6. scissor [ˈsɪzə(r)] v. 剪断
7. purification [ˌpjʊərɪfɪˈkeɪʃn] n. 提纯作用；净化作用
8. pipeline [ˈpaɪplaɪn] n. 输油管道；输气管道
9. muffler [ˈmʌflə(r)] n. 消音器
10. converter [kənˈvɜːtə(r)] n. 使发生转化的人（或物）；转换器；整流器
11. amplification [ˌæmplɪfɪˈkeɪʃn] n. 扩大；放大

Notes 注释

1. pneumatic transmission system —— 气动传动系统
2. compressed air —— 压缩空气
3. power source —— 动力源
4. mechanical equipment —— 机械设备
5. pneumatic shearing machine —— 气动剪切机
6. air compressor —— 空气压缩机
7. air cooler —— 空气冷却器
8. water separator —— 油水分离器
9. air reservoir —— 储气罐
10. air filter —— 空气过滤器
11. pressure switch —— 压力阀
12. air regulator —— 油雾器
13. feeding device —— 送料装置
14. stroke valve —— 行程阀
15. air supply device —— 气源装置
16. air motor —— 气动马达
17. flow valve —— 流量阀
18. internal lubrication —— 内部润滑
19. noise reduction —— 降噪

Reference Translation 参考译文

1. 定义

气压传动是指以压缩空气为动力源来驱动和控制各种机械设备的一种技术。随着工业机械化和自动化的发展，气动技术越来越多地应用于各个领域。

2. 工作原理

通过下面一个典型气压传动系统来介绍气动系统的工作原理。

图 8-3 所示为气动剪切机的工作原理，图示位置为剪切前的情况。空气压缩机产生的压缩空气经空气冷却器、排水分离器、储气罐、空气过滤器、减压阀、空气调节器，到达换向阀。

此时，大部分压缩空气经换向阀后进入气缸的上腔，而气缸的下腔与大气相通，气缸活塞处于最下端位置。

当上料装置把工件送入规定位置时，工件压下行程阀，使得换向阀阀芯向下运动至下端；压缩空气则进入气缸的下腔，上腔与大气相通，气缸活塞向上运动，带动剪刀上行剪断工件。

图 8-4 为用图形符号绘制的气动剪切机气压传动系统原理图。

从气动剪切机的例子可以看到，气压传动系统是将电动机的机械能转换为气体的压力能，然后通过气缸将气体的压力能转换为机械能以推动负载做功。气压传动的过程就是机械能—气压能—机械能的转换过程。

3. 组成

由上例可知，在气压传动系统中，根据功能的不同，可将气压传动系统分成四个组成部分，见表 8-3。

表 8-3 气压传动系统的组成

组成部分	元件	作用
气源装置	空气压缩机、储气罐、空气净化装置和管道等	气源装置为系统提供压缩空气
执行元件	气缸、气马达	执行元件把压缩空气的压力能转换成工作装置的机械能
控制元件	气动控制阀（如压力阀）、流量阀、换向阀和控制元件等	控制元件用来控制压缩空气的压力、流量和流动方向
辅助元件	如油雾器、消声器、转换器、显示器和传感器等	辅助元件用于元件内部润滑、消声以及信号转换、显示、放大和检测等

Part 4 Application and Features of Pneumatic System
气压系统的特点及应用

1. Advantages and Disadvantages of Pneumatic System

（1） Advantages

1） Easy to use. Pneumatic system uses air as the working medium, and air is everywhere. After use, air can be directly discharged into the atmosphere without polluting the environment.

2） Rapid response. The pneumatic system acts quickly and can achieve the required pressure and speed in a relatively short time.

Unit 8 Hydraulics and Pneumatics Transmission
液压与气压传动

3) Safety and reliability. Compressed air can be safely and reliably used in flammable, explosive, dusty, radiation and other harsh environments.

4) Easy storage. The air pressure has a high self-retaining ability, and the compressed air can be stored in the storage tank and used at any time.

Video 33

5) Long-distance transmission. Because the air flow resistance is small, and the pressure loss of air flow in the pipeline is small, the long-distance transmission of air is possible.

(2) Disadvantages

1) Velocity stability is poor. Due to the high compressibility of air, the velocity of air cylinder is sensitive to the change of load.

2) Purification and lubrication are needed. Compressed air must be well processed to remove dust and moisture. Measures must be taken to lubricate the components in the system, such as adding air regulator and other devices.

3) Small output force. The output force of the pneumatic system is small. With the same output force, the size of the pneumatic device is larger than that of the hydraulic device.

4) Noise. Pneumatic system is very noisy, so in general, mufflers need to be installed.

2. Application of Pneumatic System

Pneumatic transmission is widely used in the following areas, as shown in Table 8-4.

Table 8-4 Application of Pneumatic Transmission System

Industry	Examples
Machinery manufacturing industry	Clamping and conveying of workpieces on the mechanical production line, etc.
Electronic and electrical industry	Transporting silicon wafers, inserting and soldering components, etc.
Petroleum and chemical industry	Petroleum refining, etc.
Light industry	Alcohol, oil, gas canning, food packaging, etc.
Manipulator and industrial robot	Manipulator (Figure 8-5), assembly robots, painting robots, handling robots and welding robots, etc.

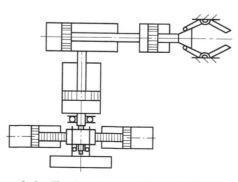

Figure 8-5 The Structure of a Pneumatic Manipulator

3. Comparison of Different Power Systems

There are four basic methods of transmitting power: mechanical, electrical, pneumatic and hydraulic power. Most applications actually use a combination of the four methods to obtain the most efficient overall system. To properly determine which method to use, it is important to know the salient features of each type. Table 8-5 lists the salient features of each type.

Table 8-5 Comparison of Different Power Systems

Property	Mechanical	Electrical	Pneumatic	Hydraulic
Energy Source	Electric motor	Water/air turbines	Pressure tank	Electric motor/Air turbine
Energy Transfer Element	Levers, gears, shafts	Electrical cables and magnetic field	Pipes and hoses	Pipes and hoses
Energy Carrier	Rigid and elastic objects	Flow of electrons	Air	Hydraulic liquids
Power-to-weight Ratio	Poor	Fair	Best	Best
Torque/Inertia	Poor	Fair	Good	Best
Stiffness	Good	Poor	Fair	Best
Response Speed	Fair	Best	Fair	Good
Dirt Sensitivity	Best	Best	Fair	Fair
Relative Cost	Best	Best	Good	Fair
Motion Type	Mainly rotary	Mainly rotary	Linear or rotary	Linear or rotary

Vocabulary 词汇

1. discharge   [dɪsˈtʃɑːdʒ]   v. 排出；放出；流出
2. pollute   [pəˈluːt]   v. 污染；弄脏
3. flammable   [ˈflæməbl]   adj. 易燃的；可燃的
4. explosive   [ɪkˈspləʊsɪv]   adj. 易爆炸的；可能引起爆炸的；易爆发的
5. dusty   [ˈdʌsti]   adj. 布满灰尘的；灰尘覆盖的
6. radiation   [ˌreɪdiˈeɪʃn]   n. 辐射；放射线
7. compressibility [kəmˌpresɪˈbɪləti]   n. 可压缩性；压缩常数
8. moisture   [ˈmɔɪstʃə(r)]   n. 潮气；水汽；水分
9. petroleum   [pəˈtrəʊliəm]   n. 石油；
10. solder   [ˈsəʊldə(r)]   v. （使）焊接，焊合
11. clamp   [klæmp]   v. 夹紧，夹住；锁住

Unit 8　Hydraulics and Pneumatics Transmission
液压与气压传动

12. turbine	['tɜːbaɪn]	n.	涡轮机；汽轮机
13. ratio	['reɪʃɪəʊ]	n.	比，比率；比例；系数
14. rigid	['rɪdʒɪd]	adj.	刚硬的，顽固的
15. elastic	[ɪ'læstɪk]	adj.	有弹力的；可伸缩的；灵活的
16. electron	[ɪ'lektrɒn]	n.	电子
17. stiffness	['stɪfnəs]	n.	刚度；硬度

Notes　注释

1. self-retaining　　　　　　　　　自保持
2. long-distance transmission　　　远距离传输
3. energy source　　　　　　　　　能量来源
4. energy transfer element　　　　 能量传递元件
5. energy carrier　　　　　　　　　能量载体
6. power-to-weight ratio　　　　　 功率和重量比
7. response speed　　　　　　　　响应速度

Reference Translation　参考译文

1. 气动系统的优缺点

（1）优点

1）使用方便。气动系统以空气作为工作介质，空气无处不在。空气用过以后可以直接排入大气，不会污染环境。

2）快速反应。气动系统动作迅速，可在较短的时间内达到所需的压力和速度。

3）安全可靠。压缩空气可安全可靠地应用于易燃、易爆、多尘埃、辐射等恶劣的环境中。

4）方便储存。气压具有较高的自保持能力，压缩空气可以储存在储气罐内，随时取用。

5）可远距离传输。由于空气流动阻力小，管道中空气流动的压力损失小，有利于远距离输送。

（2）缺点

1）速度稳定性差。由于空气可压缩性比较大，气缸的运动速度易随负载的变化而变化，稳定性较差。

2）需要净化和润滑。压缩空气必须经过良好的处理，以去除含有的灰尘和水分。系统中必须采取措施对元件进行润滑，如加装油雾器等装置。

3）输出力小。气动系统输出的力小，在输出力相同的情况下，气动装置比液压装置尺寸要大得多。

4）噪声大。气动系统排放空气的声音很大，因此气动系统一般需要加装消音器。

2. 气动系统的应用

气压传动系统可应用在诸多领域，见表8-4。

表 8-4 气压传动系统的应用

行业名称	应用场合举例
机械制造业	机械生产线上工件的装夹及搬送
电子及电气行业	用于硅片的搬运，元器件的插装与锡焊等
石油、化工业	如石油提炼加工等
轻工业	例如酒精类、油类、煤气罐装，以及各种食品的包装等
机械臂和工业机器人	机械臂（见图 8-5）、装配机器人、喷漆机器人、搬运机器人以及焊接机器人等

3. 不同传动系统的比较

传递动力的基本方法有四种：机械传动、电气传动、气压传动和液压传动。大多数应用程序实际上使用四种方法的组合来获得最有效的整体系统。为了正确确定使用哪种方法，了解每种类型的显著特征是很重要的。表 8-5 列出了每种类型的显著特征。

表 8-5 不同传动系统的对比

属性	机械传动	电气传动	气动传动	液压传动
输入动力源	电动机	水涡轮/空气涡轮	压力罐	电动机/空气涡轮
能量输送元件	杠杆、齿轮、轴	电缆以及磁场	管道和软管	管道和软管
能量传输介质	刚性和弹性物体	电子流	空气	液体
功率重量比	不好	一般	很好	很好
转矩	不好	一般	好	很好
刚度	好	不好	一般	很好
响应速度	一般	很好	一般	好
对污垢敏感性	很好	很好	一般	一般
相对成本	很好	很好	好	一般
运动类型	主要是旋转	主要是旋转	线性或者旋转	线性或者旋转

Unit 9　Motor Drive

电动机拖动

DC Motor
直流电动机

1. Definition

A DC motor is a class of rotary electrical machines that converts direct current(DC) electrical energy into mechanical energy. DC motors were the first form of motors widely used, as they could be powered from existing direct current lighting power distribution systems. A DC motor's speed can be controlled over a wide range, using either a variable supply voltage or by changing the strength of current in its field windings.

Video 34

2. Structure

A DC motor mainly composes of two parts, the rotatable part and the stationary part.

The rotatable part is called the rotor or armature, and the stationary part is called the stator. The gap between stator and rotor is called air gap.

The stator includes main magnetic pole, commutation pole, brush device, machine base, end cover and bearing, and the rotor includes armature core, armature winding, commutator, rotating shaft, fan and bracket.

The structure of a DC motor is shown in Figure 9-1.

3. Principle

Figure 9-2 illustrates a simple, two-pole, brushed, DC motor. It converts DC electric energy into mechanical energy and drives the machinery on the shaft. When the rotor moves to the position shown in Figure 9-2a, the ab conductor is just below the N pole and the cd conductor is below the S

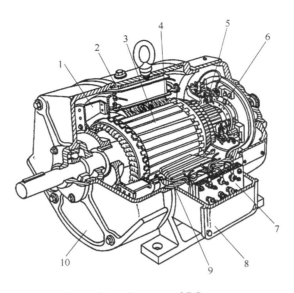

Figure 9-1 Structure of DC motor
1—Fan 2—Engine base 3—Armature 4—Main magnetic pole 5—Brush device
6—Commutator 7—Terminal block 8—Outlet box 9 Communication pole 10—End cover

pole. The DC current flows into the armature winding via brush A. The current direction is shown by the arrow in Figure 9-2a. The direction of force acting on the conductor is determined by the Left Hand Rule. The result shows that the torque produced by the force imposed on the conductor ab and cd is counterclockwise, and the motor rotates counterclockwise.

When the rotor rotates 180 degrees to the position shown in Figure 9-2b the conductor cd is below the N pole and the conductor ab is below the S pole. The current flows into the armature through the A brush. The current direction is shown by the arrow in Figure 9-2b. According to the Left Hand Rule, it can be determined that the torque produced by the force imposed on the conductors ad and cd is counterclockwise.

It can be seen that although the internal current direction of the conductor changes, the direction of torque generated by the force remains unchanged, and the rotor rotates continuously in the same direction.

(1) Torque and Speed of a DC Motor

1) Counter electromotive force (EMF) equation. Counter EMF is the electromotive force or "voltage" that opposes the change in current which induced it. When we apply a voltage, and the motor begins to spin, it will generate a voltage that opposes the external voltage we apply to it, i.e. counter EMF. The DC motor's counter EMF is proportional to the product of the machine's total flux strength and armature's speed:

$$E_b = K_b \varphi n$$

Where E_b is the counter EMF(V), K_b is counter EMF equation constant, φ is machine's total flux(Wb), and n is armature frequency(r/min).

Unit 9 Motor Drive
电动机拖动

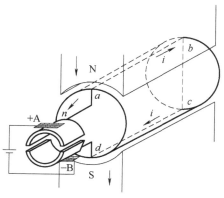 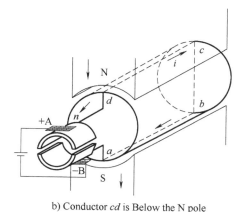

a) Conductor *ab* is Below the N pole   b) Conductor *cd* is Below the N pole

Figure 9-2 Principle of a Simple DC Motor

2) Voltage balance equation. The DC motor's input voltage must overcome the counter EMF as well as the voltage drop created by the armature current across the motor resistance. Motor resistance refers to the combined resistance across the brushes, armature winding and series field winding:

$$U_m = E_b - I_a R_m$$

Where U_m is motor input voltage(V), E_b is the counter EMF(V), R_m is motor resistance(Ω), I_a is armature current(A).

3) Torque equation. Torque is the force with which the motor spins. The DC motor's torque is proportional to the product of the armature current and the machine's total flux strength:

$$T = K_T I_a \varphi$$

where T is motor torque(Nm), φ is machine's total flux(Wb), I_a is armature current(A), K_T is torque equation constant.

(2) DC motor starting When power is first applied to a motor, the armature does not rotate, due to the mechanical inertia of the rotating body. At that instant the counter EMF is zero and the only factor limiting the armature current is the armature resistance and inductance. Usually the armature resistance of a motor is very small, therefore the current through the armature would be very large in the starting stage.

This current will disturb the power grid, generate mechanical impact on the unit and spark the commutator. That is the reason why some precautionary measures are taken during the starting of the DC motors.

There are two methods to limit the starting current: one is to decrease the supply voltage U; the other is to increase the resistance of armature circuit.

1) Reduce supply voltage starting.

Figure 9-3 is the wiring diagram for reducing supply voltage starting. When starting, the excitation winding is connected to the power supply, and the voltage of armature circuit is adjusted from low to high.

2) Variable resistance starting.

The variable resistance starting is shown in Figure 9-4. In the start-up process, the start-up resistance in series with the armature circuit is removed stage by stage.

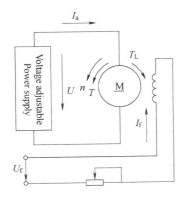

Figure 9-3 Reduce Supply Voltage Starting   Figure 9-4 Variable Resistance Starting

(3) Speed control of DC motor

The speed of a DC motor is equal to:

$$n = \frac{U_m + I_m R_q}{K_D \varphi}$$

Where U_m is motor input voltage(V), R_m is motor resistance(Ω), I_a is armature current(A), K_b is counter EMF equation constant, and φ is machine's total flux(Wb).

Therefore, speed of DC motors can be controlled by changing the following variables:

1) The terminal voltage of the armature, and the corresponding mechanical characteristic is shown in Figure 9-5a.

2) The external resistance in armature circuit, and the corresponding mechanical characteristic is shown in Figure 9-5b.

3) The flux per pole, and the corresponding mechanical characteristic is shown in Figure 9-5c.

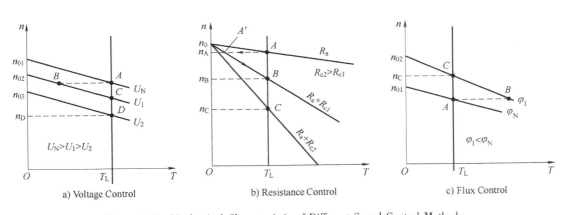

Figure 9-5 Mechanical Characteristic of Different Speed Control Methods

Unit 9 Motor Drive
电动机拖动

Vocabulary　词汇

1. winding	['waɪndɪŋ]	n.	绕，缠；线圈
2. pole	[pəʊl]	n.	[物] 极点，顶点
3. illustrate	['ɪləstreɪt]	v.	说明；表明；给……加插图
4. armature	['ɑːmətʃə(r)]	n.	电枢
5. commutator	['kɒmjuteɪtə(r)]	n.	换向器，转接器
6. counter	['kaʊntə(r)]	adj.	相反的
7. proportional	[prə'pɔːʃnl]	adj.	比例的，成比例的
8. reverse	[rɪ'vɜːs]	v.	（使）反转
9. frequency	['friːkwənsi]	n.	频率，次数
10. permanent	['pɜːmənənt]	adj.	永久（性）的
11. magnet	['mæɡnət]	n.	磁铁；磁体
12. excitation	[,eksɪ'teɪʃn]	n.	励磁
13. series	['sɪəriːz]	n.	串联；
14. shunt	[ʃʌnt]	n.	转轨；分流器
15. compound	['kɒmpaʊnd]	n.	复合物；复合词
16. resistance	[rɪ'zɪstəns]	n.	电阻
17. permissible	[pə'mɪsəbl]	adj.	可允许的；许可的
18. precautionary	[prɪ'kɔːʃənəri]	adj.	预先警戒的，小心的
19. bracket	['brækɪt]	n.	支架

Notes　注释

1. field winding　　　　　　　　　　励磁绕组
2. Fleming's Left Hand Rule　　　　弗莱明左手定则
3. counter electromotive force　　　反电动势
4. DC motor　　　　　　　　　　　直流电动机
5. rotary electrical machine　　　　 旋转电动机
6. direct current（DC）　　　　　　直流电
7. lighting power distribution system　照明配电系统
8. variable supply voltage　　　　　可变电源电压
9. air gap　　　　　　　　　　　　气隙
10. magnetic pole　　　　　　　　　磁极
11. commutation pole　　　　　　　换向极
12. brush device　　　　　　　　　电刷装置
13. machine base　　　　　　　　　基座
14. armature core　　　　　　　　　电枢铁心
15. armature winding　　　　　　　电枢绕组
16. rotating shaft　　　　　　　　　旋转轴

17. voltage balance equation　　　　电压平衡方程
18. torque equation　　　　　　　　　转矩方程
19. armature current　　　　　　　　 电枢电流
20. reduce supply voltage starting　　降电压起动
21. variable resistance starting　　　　变电阻起动

Reference Translation　参考译文

1. 定义

直流电动机是一类能将直流电能转换为机械能的旋转电机。直流电动机是第一种广泛使用的电动机，因为它们可以由现有的直流照明配电系统供电。直流电动机的速度可以在很宽的范围内控制，既可以通过改变电压来控制，也可以通过改变其磁场绕组中的电流来控制。

2. 结构

直流电动机在结构上主要由旋转部分和静止部分组成。

旋转部分称为转子或电枢，静止部分称为定子。定子与转子之间有间隙，称为气隙。

定子部分包括主磁极、换向极、电刷装置、机座、端盖和轴承等部件；转子部分包括电枢铁心、电枢绕组、换向器、转轴、风扇和支架等部件。

直流电动机的结构如图 9-1 所示。

3. 原理

图 9-2 所示为一个简单的两极有刷直流电动机。它把直流电能转换成机械能，带动轴上的生产机械。当电动机转子转到图 9-2a 所示位置时，ab 导体刚好在 N 极下，cd 导体在 S 极下。直流电流经电刷 A 流入电枢绕组，电流方向如图 9-2b 中箭头所示。导体受力方向由左手定则判定，判定结果导体 ab 和 cd 受力产生的转矩均为逆时针方向，电动机逆时针旋转。

当电动机转子转过 180°时转到图 9-2b 所示位置时，导体 cd 在 N 极下，导体 ab 在 S 极下。电流经电刷 A 由 d 端流入线圈。电流方向如图 9-2b 中箭头所示。根据左手定则可判定 ad 和导体 cd 受力产生的转矩为逆时针方向。

由此可知，虽然导体内部电流方向变了，但受力产生的转矩方向不变，转子连续旋转方向不变。

（1）直流电动机的转矩和转速

1）反电动势（EMF）方程：反电动势是与感应电流的变化相反的电动势。当施加一个电压，电动机开始旋转时，它会产生一个与施加的外部电压相反的电压，即反电动势。直流电动机的反电动势与电动机总磁通强度和电枢速度的乘积成正比：

$$E_b = K_b \varphi n$$

式中，E_b 为反电动势（V），K_b 为反电动势方程常数，φ 为电动机总磁通量（Wb），n 为电枢频率（r/min）。

2）电压平衡方程：直流电动机的输入电压必须克服反电动势以及电枢电流在电动机电阻上产生的电压降。电动机电阻是指电刷、电枢绕组和串联磁场绕组的组合电阻，则有：

$$U_m = E_b - I_a R_m$$

式中，U_m 为电动机输入电压（V），E_b 为反电动势（V），R_m 为电动机电阻（Ω），I_a 为电枢电流（A）。

3）转矩方程：

转矩是使电动机旋转的力。直流电动机的转矩与电枢电流和电动机总磁通强度的乘积成正比：

$$T = K_T I_a \varphi$$

式中，T 为电动机转矩（Nm），φ 为电动机总磁通量（Wb），I_a 为电枢电流（A），K_T 为转矩方程常数。

（2）直流电动机起动　当电动机接通电源后，由于旋转体的机械惯性，电枢不转动。此时反电动势为零，限制电枢电流的唯一因素是电枢电阻和电枢电感。通常电动机的电枢电阻很小，因此在起动阶段通过电枢的电流会很大。

这一电流会使电网受到扰动、机组受到机械冲击、换向器发生火花。这就是为什么在直流电动机起动过程中要采取一些预防措施的原因。

限制起动电流的措施有两个：一是降低电源电压 U；二是加大电枢回路电阻。

1）降低电源电压起动。图 9-3 是降低电源电压起动时的接线图。起动时，先将励磁绕组接通电源，然后从低向高调节电枢回路的电压。

2）电枢串电阻起动。电枢串电阻起动如图 9-4 所示。在起动过程中，将串入电枢回路的起动电阻分级切除。

（3）直流电动机速度控制

直流电动机的转速等于：

$$n = \frac{U_m + I_m R_a}{K_b \varphi}$$

式中，U_m 为电动机输入电压（V），R_m 为电动机电阻（Ω），I_a 为电枢电流（A），K_b 为反电动势方程常数，φ 为电动机总磁通量（Wb）。

直流电动机的转速可以通过改变以下变量来实现：

1）改变电枢的端电压，对应的机械特性如图 9-5a 所示。
2）改变电枢电路的外电阻，对应的机械特性如图 9-5b 所示。
3）改变每极磁通量，对应的机械特性如图 9-5c 所示。

Part 2　Induction Motor　交流异步电动机

1. Definition

Induction motor, or asynchronous motor, is an electric drive device which converts electric energy into mechanical energy. When the stator winding is connected to three-phase AC power supply, a rotating magnetic field is created which will get the rotor to start turning. Induction motors are generally divided into two categories, single-phase AC asynchronous motors and three-phase AC asynchronous motors.

Video 35

Three-phase AC asynchronous motor has the advantages of simple structure, reliable operation, low price, strong overload capacity, convenient use, installation and maintenance, and is widely used in various fields.

2. Structure

Three-phase AC asynchronous motor is mainly composed of stator and rotor. There is a small gap between stator and rotor. Based on the rotor structure, three-phase AC asynchronous motor can be divided into squirrel cage type and winding type. The structure of three-phase winding asynchronous motor is shown in Figure 9-6, and the structure of three-phase squirrel cage asynchronous motor is shown in Figure 9-7.

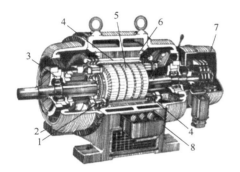

Figure 9-6 Three-Phase Winding
Asynchronous Motor
1—Rotor winding 2—End cap 3—Bearing
4—Stator winding 5—Rotor 6—Stator
7—Slip ring 8—Outlet box

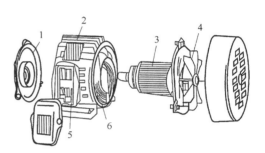

Figure 9-7 Three-Phase Squirrel Cage
Asynchronous Motor
1—End cap 2—Stator 3—Rotor
4—Wind wing 5—Connection box
6—Stator winding

3. Principle

In induction motors, the AC power supplied to the motor's stator creates a magnetic field that rotates. The rotating magnetic field induces currents in the windings of the rotor. The induced currents in the rotor windings in turn create magnetic fields in the rotor that react against the stator field. Figure 9-8 shows the rotating magnetic field set up in the three-phase induction motor.

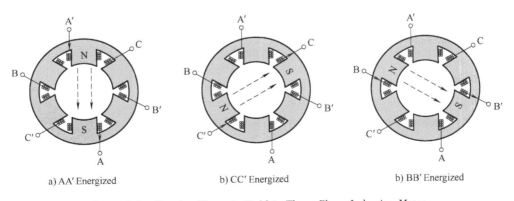

a) AA' Energized   b) CC' Energized   b) BB' Energized

Figure 9-8 Rotating Magnetic Field in Three-Phase Induction Motor

Due to Lenz's Law, the direction of the magnetic field created will be such as to oppose the change in current through the rotor windings. The cause of induced current in the rotor windings is the rotating stator magnetic field, so to oppose the change in rotor-winding currents the rotor will start to rotate in the direction of the rotating stator magnetic field. The rotor accelerates until the magnitude of induced rotor current and torque balances the applied mechanical load on the rotation of the rotor.

Since rotation at synchronous speed would result in no induced rotor current, an induction motor always operates slightly slower than synchronous speed.

4. Synchronous Speed

An AC motor's synchronous speed, n_s, is the rotation rate of the stator's magnetic field:

$$n_s = 60f/p$$

Where f is the frequency of the power supply, p is the number of magnetic poles, and n_s is the synchronous speed of the machine.

5. Slip

Slip s is defined as the difference between synchronous speed and operating speed, at the same frequency, expressed in rpm, or in percentage or ratio of synchronous speed.

$$s = \frac{n_s + n_r}{n_s}$$

Where n_s is stator electrical speed, n_r is rotor mechanical speed. Slip, which varies from zero at synchronous speed and 1 when the rotor is at rest, determines the motor's torque. Since the short-circuited rotor windings have small resistance, even a small slip induces a large current in the rotor and produces significant torque.

6. Starting of Three-Phase Induction Motor

A three-phase induction motor is self-starting. The purpose of a starter is not to just start the motor, but it performs the two main functions:

1) To decrease the starting current.
2) To provide overload and under voltage protection.

The three-phase induction motor may be started by connecting the motor directly to the full voltage of the supply. The motor can also be started by applying a reduced voltage to the motor. The torque of the induction motor is proportional to the square of the applied voltage. Thus, a greater torque is exerted by a motor when it is started on full voltage than when it is started on the reduced voltage.

Vocabulary 词汇

1. stator   ['steɪtə]   n. 定子
2. rotor   ['rəʊtə(r)]   n. 转子
3. synchronous   ['sɪŋkrənəs]   adj. 同步的
4. slip   [slɪp]   n. 转差率
5. exert   [ɪɡ'zɜːt]   v. 发挥；运用

6. bearing　　　　　　　['beəriŋ]　　　　　　n. 轴承

Notes　注释

1. induction motor　　　　　　　　　　　感应电动机
2. asynchronous motor　　　　　　　　　异步电动机
3. stator winding　　　　　　　　　　　　定子绕组
4. three-phase AC power supply　　　　 三相电源
5. rotating magnetic field　　　　　　　旋转磁场
6. Lenz's Law　　　　　　　　　　　　　楞次定律
7. single-phase AC asynchronous motor　单相异步电动机
8. three-phase AC asynchronous motor　 三相异步电动机
9. three-phase squirrel cage asynchronous motor　三相笼型异步电动机
10. rotor winding　　　　　　　　　　　转子绕组
11. end cap　　　　　　　　　　　　　　端盖
12. slip ring　　　　　　　　　　　　　　集电环
13. outlet box　　　　　　　　　　　　　出线盒
14. connecting box　　　　　　　　　　　接线盒
15. wind wing　　　　　　　　　　　　　风翼
16. stator magnetic field　　　　　　　　定子磁场
17. synchronous speed　　　　　　　　　同步转速
18. self-starting　　　　　　　　　　　　自起动

Reference　参考译文

1. 定义

交流异步电动机，或称异步电动机，是一种将电能转化为机械能的电力拖动装置。对定子绕组通三相交流电源后，产生旋转磁场并使转子转动。交流异步电动机一般分为两类，单相交流异步电动机和三相交流异步电动机。

三相交流异步电动机具有结构简单、运行可靠、价格便宜、过载能力强，以及使用、安装、维护方便等优点，被广泛应用于各个领域。

2. 结构

三相交流异步电动机主要由定子和转子两大部分组成，定子和转子之间有很小的气隙。按转子结构的不同，三相异步电动机分为笼型和绕线转子两大类。图9-6所示为三相绕线转子异步电动机的结构，图9-7所示为三相笼型异步电动机的结构。

3. 原理

在交流异步电动机中，供给电动机定子的交流电源产生一个旋转的磁场。旋转磁场在转子绕组中产生感应电流。转子绕组中的感应电流反过来在转子中产生磁场，与定子磁场发生反作用。图9-8所示为三相交流异步电动机中产生的旋转磁场。

根据楞次定律，在转子绕组中产生的磁场方向将对抗转子绕组的感应电流变化。因为转子绕组中产生感应电流的原因是旋转的定子磁场，因此为了对抗转子绕组电流的变化，转子将开始随着定子磁场的方向旋转。转子持续加速直到转子感应电流和转矩与转子旋转时施加的机械负载平衡。

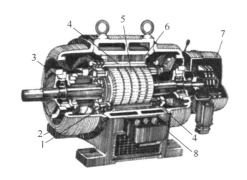

图9-6 三相绕线转子异步电动机的结构
1—转子绕组 2—端盖 3—轴承 4—定子绕组 5—转子
6—定子 7—集电环 8—出线盒

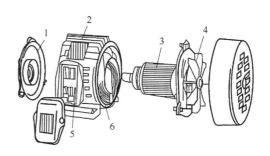

图9-7 三相笼型异步电动机的结构
1—端盖 2—定子 3—转子 4—风翼
5—接线盒 6—定子绕组

由于同步转速下的旋转不会产生转子感应电流,因此交流异步电动机的运行速度总是略慢于同步转速。

4. 同步速度

交流异步电动机的同步速度 n_s 是定子磁场的旋转速度:

$$n_s = 60f/p$$

式中,f 是电源频率,p 是磁极数,n_s 是电动机的同步转速。

5. 转差率

转差率 s 是指同步转速与工作转速在同一频率下的差,或用同步转速的百分比或比率表示,即

$$s = \frac{n_s - n_r}{n_s}$$

其中,n_s 为定子电气转速,n_r 为转子机械转速。在同步转速下,转差率为0,在转子静止时,转差率为1。转差率决定电动机的转矩。由于短路的转子绕组电阻很小,即使是很小的转差率也会在转子中产生很大的电流并产生很大的转矩。

6. 三相异步交流电机的起动

三相异步交流电动机是自起动的。加入起动器的目的不只是起动电动机,而是执行两个主要功能:

1) 降低起动电流。
2) 提供过载和欠电压保护。

三相异步交流电动机可以通过将电动机直接连接到电源的全电压来起动,也可以通过向电动机施加降低的电压来起动电动机。三相异步交流电动机的转矩与施加电压的二次方成正比。因此,当电动机在全电压下起动时,其施加的转矩大于在降低电压下起动时的转矩。

Part 3 Stepper Motor
步进电动机

1. Definition

Stepper motor is an open-loop driving element that converts electrical pulse signal into angular or linear displacement.

Stepper motors are classified into permanent magnet stepper motor, variable reluctance stepper motor and hybrid synchronous stepper motor. Among them, hybrid synchronous stepper motors combine the advantages of permanent magnet and variable reluctance stepper motors, and are most widely used.

2. Structure

The basic structure of three-phase variable reluctance stepping motor is shown in Figure 9-9. Stator windings are coils wound on six uniformly distributed teeth on the stator core. The coils on the two opposite teeth in diameter direction are connected in series to form one-phase. If any phase is electrified, a set of N and S poles will be formed (the direction is shown in Figure 9-9). Forty teeth are evenly distributed on the rotor, and five small teeth are distributed on each tooth of the stator core.

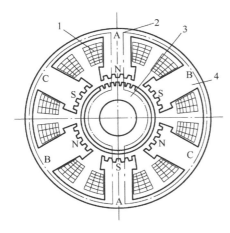

Figure 9-9 Basic Structure of Three-Phase Variable Reluctance Stepper Motor

1—Stator windings 2—Pole A 3—Rotor 4—Stator core

Video 36

3. Principle

When phase A is energized, the teeth on the rotor are aligned with the teeth on the stator AA. If phase A is cut off and phase B is electrified, under the action of magnetic force, the teeth of the rotor are aligned with the small teeth on the stator BB, and the rotor rotates clockwise by 3 degrees. If the control circuit continuously controls the turn-on and power-off of the control winding in the order of

A→B→C→A→..., the rotor of the stepper motor rotates clockwise. If the electrification sequence is changed to A→C→B→A→..., the rotor of the stepper motor rotates counter-clockwise.

4. Step Angle in Stepper Motor

Step angle is defined as the angle which the rotor of a stepper motor moves when one pulse is applied to the input of the stator. The step angle may be found from the following formula:

$$S_A = \frac{360°}{S_{PR}}$$

where S_A = step angle in degrees, S_{PR} = steps per revolution

The resolution or the step number of a motor is the number of steps it makes in one revolution of the rotor. The smaller the step angle, the higher the resolution of the stepper motor will be. The accuracy of positioning of the objects by the motor depends on the resolution. A standard motor will have a step angle of 1.8 degrees with 200 steps per revolution. The various step angles like 90, 45 and 15 degrees are common in simple motors.

5. Stepper Motor Drive

Each phase of stepper motor is electrified according to certain rules. Special drivers are needed to cooperate with the controller to complete the operation.

Stepper motor drives are generally composed of a ring distributor and a power amplifier, as shown in Figure 9-10. The block diagram of stepper motor and drive is shown in Figure 9-11.

a) Stepper Motor   b) Stepper Motor Drive

Figure 9-10 Stepper Motor and Drive

Ring distributor distributes electric pulse according to the mode of electrification, and power amplifier amplifies the small output signal of the ring distributor.

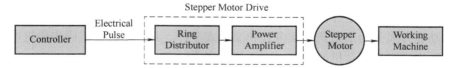

Figure 9-11 Block Diagram of Stepper Motor and Drive

Vocabulary 词汇

1. angular　　　　　['æŋgjələ(r)]　　　　adj. 有角的；用角测量的
2. linear　　　　　 ['lɪniə(r)]　　　　　　adj. 直线的，线性的
3. hybrid　　　　　['haɪbrɪd]　　　　　　adj. 混合的；杂种的
4. reluctance　　　 [rɪ'lʌktəns]　　　　　n. 磁阻
5. clockwise　　　 ['klɒkwaɪz]　　　　　adv. 顺时针方向转动地
6. formula　　　　 ['fɔːmjələ]　　　　　 n. 公式，准则

7. diameter	[daɪˈæmɪtə(r)]	n. 直径，直径长
8. magnetic	[mægˈnetɪk]	adj. 有磁性的
9. align	[əˈlaɪn]	v. 使成一线，整齐
10. revolution	[ˌrevəˈluːʃn]	n. 旋转
11. resolution	[ˌrezəˈluːʃn]	n. 分辨率
12. distributor	[dɪˈstrɪbjətə(r)]	n. 分配者；[电] 配电盘
13. electrified	[ɪˈlektrɪfaɪd]	adj. 通电的；带电的　v. 使电气化；使通电
14. energized	[ˈenədʒaɪzd]	v. 使通电

Notes　注释

1. stepper motor　　　　　　　　　步进电动机
2. permanent magnet stepper motor　　永磁式步进电动机
3. variable reluctance stepper motor　　可变磁阻步进电动机（反应式步进电动机）
4. hybrid synchronous stepper motor　　混合同步步进电动机
5. counter-clockwise　　　　　　　　逆时针转动的
6. stator core　　　　　　　　　　　定子铁心
7. turn-on　　　　　　　　　　　　打开电源
8. power-off　　　　　　　　　　　关闭电源
9. step angle　　　　　　　　　　　步进角
10. stepper motor drive　　　　　　　步进电动机驱动器
11. ring distributor　　　　　　　　　环形分配器
12. power amplifier　　　　　　　　　功率放大器
13. electric pulse　　　　　　　　　　电脉冲

Reference Translation　参考译文

1. 定义

步进电动机是一种将电脉冲信号转变为相应的角位移或线位移的开环控制驱动元件。步进电动机按励磁方式分为永磁式、反应式和混合式三种。其中，混合式同步步进电动机综合了永磁式和反应式步进电动机的优点，应用最为广泛。

2. 基本结构

三相反应式步进电动机的基本结构如图9-9所示。定子绕组是绕在定子铁心上的6个均匀分布的齿上的线圈，直径方向上相对的两个齿上的线圈串联在一起，构成一相控制绕组。若任一相绕组通电，便形成一组N、S磁极（方向如图所示）。转子上均匀分布40个齿，而定子铁心每个齿上又开了5个小齿。

3. 工作原理

当A相绕组通电时，转子上的齿与定子AA上的小齿对齐。若A相断电，B相通电，在磁力作用下，转子的齿与定子BB上的小齿对齐，转子沿顺时针方向转过3°。如果控制电路不断地按A→B→C→A→…的顺序控制控制绕组的通、断电，步进电动机的转子则不停地顺时针转动。若通电顺序改为A→C→B→A→…，则实现逆时针转动。

4. 步进电动机的步距角

步距角是指当一个脉冲作用于定子的输入端时，步进电动机的转子移动的角度。步进角可由以下公式得出：

$$S_A = \frac{360°}{S_{PR}}$$

式中，S_A = 步距角（单位：度），S_{PR} = 每转步数。

电动机的分辨率是其在转子旋转一圈内所走的步数。步矩角越小，步进电动机定位的分辨率越高。电动机对物体定位的精度取决于分辨率。标准电动机的步距角为 1.8°，每转 200 步。90°、45°和 15°等各种步矩角在简单电动机中很常见。

5. 步进驱动器

步进电动机每相绕组按照一定的规律轮流通电，需要专门的驱动器，配合控制器完成操作。

步进驱动器一般由环形分配器和功率放大器组成，如图 9-10 所示。步进电动机和驱动器的框图如图 9-11 所示。

环形分配器是将电脉冲按通电方式进行分配；功率放大器是将环形分配器输出的小信号进行功率放大。

Part 4 Servo Motor
伺服电动机

1. Definition

A servo motor is a rotary actuator or linear actuator that allows for precise control of angular or linear position, velocity and acceleration. It consists of a suitable motor coupled to a sensor for position feedback. It also requires a relatively sophisticated controller, often a dedicated module designed specifically for use with servo motors.

Video 37

Servo motors can be divided into DC servo motors and AC servo motors according to the types of current.

2. DC Servo Motor

(1) Basic structure There is no essential difference between the basic structure of DC servo motors and ordinary DC motors. According to the different excitation modes, it can be divided into two types: separate excitation and permanent magnet excitation.

(2) Working principle The connection circuit of a DC servo motor is shown in Figure 9-12. The excitation winding is connected to a constant voltage DC power supply. The armature winding is connected to the control voltage as the control winding. When the control voltage is changed, the DC servo motor is

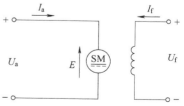

Figure 9-12 The Connection Circuit of a DC Servomotor

in the state of speed regulation. Its mechanical characteristics are a series of parallel straight lines, as shown in Figure 9-13. The regulation characteristics of DC servo motors are shown in Figure 9-14.

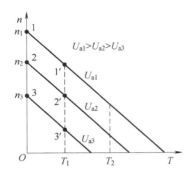

Figure 9-13　Mechanical Characteristics of a DC Servo Motor

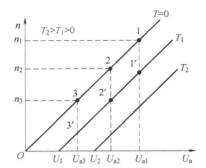

Figure 9-14　Regulation Characteristics of a DC Servo motor

3. AC Servo Motor

(1) Basic structure　At present, the servo motors used in industrial robots are generally synchronous AC servo motors, whose motor body employs permanent magnet synchronous motor (PMSM). PMSM consists of stator and rotor, as shown in Figure 9-15.

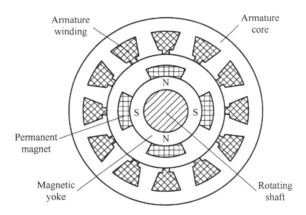

Figure 9-15　Synchronous AC Servo Motor

(2) Working principle　When symmetrical three-phase current is applied in armature winding of PMSM, the stator will generate a rotating magnetic field with synchronous speed. In the steady state, the speed of the rotor is the synchronous speed of the magnetic field. Therefore, the rotating magnetic field of the stator and the main pole magnetic field generated by the permanent magnet of the rotor remain static, and they interact with each other to generate electromagnetic torque, driving the rotor to rotate.

When the load changes, the instantaneous speed of the rotor will change. At this time, the speed and position of the rotor are detected by the sensor. Based on the position of the permanent magnet field of the rotor, the current, phase and frequency in the stator winding are controlled by the inverter, and the continuous torque will be produced on the rotor. This is the principle of closed-loop control of permanent magnet synchronous motor.

4. Servo Drive

Servo drive is a kind of controller used to control servo motor, as shown in Figure 9-16.

Figure 9-16 Servo Motors and Drives

Servo drive usually controls the servo motor by position, speed and moment to achieve high-precision positioning of the transmission system.

Vocabulary 词汇

1. servo   ['sɜːvəʊ]   n. （机器的）伺服传动装置，伺服系统
2. inverter   [ɪnˈvɜːtə]   n. 逆变器

Notes 注释

1. rotary actuator   旋转执行器
2. DC servo motor   直流伺服电动机
3. AC servo motor   交流伺服电动机
4. separate excitation   他励
5. permanent magnet excitation   永磁式
6. excitation winding   励磁绕组
7. armature winding   电枢绕组
8. mechanical characteristic   机械特性
9. regulation characteristic   调节特性
10. permanent magnet synchronous motor   永磁同步电动机
11. electromagnetic torque   电磁转矩
12. symmetrical three-phase current   对称三相电流
13. rotating magnetic field   旋转磁场
14. servo drive   伺服驱动器
15. servo controller   伺服控制器
16. servo amplifier   伺服放大器

17. transmission system　　　　　　　　传动系统

Reference Translation　参考译文

1. 定义

伺服电动机是一种旋转执行器或线性执行器，允许精确控制角度或线性位置、速度和加速度。它包括一个合适的电动机以及一个耦合的位置反馈传感器。它还需要一个相对复杂的控制器，通常是专门为伺服电动机设计的专用模块。

伺服电动机按电流种类的不同，可分为直流伺服电动机和交流伺服电动机两大类。

2. 直流伺服电动机

（1）基本结构　一般的直流伺服电动机的基本结构与普通直流电动机并无本质的区别。按励磁方式的不同，直流伺服电动机可分为他励式和永磁式两种。

（2）工作原理　直流伺服电动机的接线如图9-12所示。励磁绕组连接在一个电压恒定的直流电源上。电枢绕组作为控制绕组接到控制电压上。改变控制电压的数值，直流伺服电动机则处于调速状态，它的机械特性是一组平行直线，如图9-13所示。直流伺服电动机的调节特性如图9-14所示。

3. 交流伺服电动机

（1）基本结构　目前，工业机器人采用的伺服电动机一般为同步交流伺服电动机，其本体为永磁同步电动机。永磁同步电动机由定子和转子两部分构成，如图9-15所示。

（2）工作原理　当永磁同步电动机的电枢绕组中通过对称的三相电流时，定子将产生一个以同步转速工作的旋转磁场。在稳态下，转子的转速恒为磁场的同步转速。于是，定子旋转磁场与转子永磁体产生的主极磁场保持静止，它们之间相互作用，产生电磁转矩，拖动转子旋转。

当负载发生变化时，转子的瞬时转速就会发生变化，这时，如果通过检测传感器检测转子的速度和位置，根据转子永磁体磁场的位置，利用逆变器控制定子绕组中的电流大小、相位和频率，便会产生连续的转矩作用在转子上，这就是闭环控制的永磁同步电动机的工作原理。

4. 伺服驱动器

伺服驱动器是用来控制伺服电动机的一种控制器，如图9-16所示。伺服驱动器一般通过位置、速度和力矩三种方式对伺服电动机进行控制，实现高精度的传动系统定位。

Unit 10 Microcomputer and Microprocessor

Part 1 Microprocessor Technology
微处理器技术

1. Definition

A microprocessor is a computer processor that incorporates the functions of a central processing unit on a single or a few integrated circuits(IC). The microprocessor is a multipurpose, clock driven, register based, digital integrated circuit. The microprocessor accepts binary data as input, processes it according to instructions, and provides results as output.

Video 38

2. Architecture of Microprocessor

The microprocessor reads each instruction from the memory, decodes it and executes it. The processing is in the form of arithmetic and logical operations. As per instruction, the data is retrieved from memory or taken from an input device and the result of processing is stored in the memory or delivered to an appropriate output device. To perform all these functions, the microprocessor incorporates various functional units in an appropriate manner. Such an internal structure of microprocessor is known as its architecture.

A typical microprocessor architecture is shown in Figure 10-1.

1) Arithmetic-logic unit(ALU): performs addition, subtraction, and operations such as AND or OR.

2) Control unit: retrieves instruction codes from memory and initiates the sequence of operations required for the ALU to carry out the instruction.

Figure 10-1 Architecture of Microprocessor

3) Buses: through the address bus, the microprocessor transmits the address of that I/O device or memory locations which it desires to access. The data bus is used by the microprocessor to send and receive data and instructions to and from different devices(I/O and memory). The control bus is used for transmitting and receiving control signals between the microprocessor and various devices in the system.

4) Internal registers: a number of registers are normally included in the microprocessor. These are used for temporary storage of data, instructions and addresses during execution of a program.

5) Input/Output(I/O) devices are used for data input/output. Input devices are used for feeding data into the microprocessor, and the output devices are used for delivering the results of computations to the outside world.

6) Memory stores both the instructions to be executed and the data involved. It usually consists of both RAMs(random-access memories) and ROMS(read-only memories).

3. Application of Microprocessor

Thousands of items that were traditionally not computer-related include microprocessors. Figure 10-2 shows some applications of microprocessor.

Figure 10-2 Applications of Microprocessors

The Raspberry Pi is an interesting application of microprocessors. It is a small single-board computer developed by the Raspberry Pi Foundation to promote teaching of basic computer science in

Unit 10 Microcomputer and Microprocessor
微机和微处理器

schools and in developing countries. The original model became far more popular than anticipated, selling outside its target market for uses such as robotics. Figure 10-3 shows a Raspberry Pi.

Figure 10-3 A Raspberry Pi

Vocabulary 词汇

1. microprocessor   [ˌmaɪkrəʊˈprəʊsesə(r)]   n. 微处理器
2. circuit   [ˈsɜːkɪt]   n. 电路，线路
3. multipurpose   [ˌmʌltiˈpɜːpəs]   adj. 多种用途的
4. register   [ˈredʒɪstə(r)]   n. 寄存器
5. binary   [ˈbaɪnəri]   adj. 二进位的；n. 二进制
6. arithmetic   [əˈrɪθmətɪk]   adj. 算术的
7. architecture   [ˈɑːkɪtektʃə(r)]   n. 计算机构造
8. bus   [bʌs]   n. 总线
9. involve   [ɪnˈvɒlv]   v. 包含；使参与，牵涉
10. raspberry   [ˈrɑːzbəri]   n. 树莓
11. foundation   [faʊnˈdeɪʃn]   n. 基金会
12. anticipate   [ænˈtɪsɪpeɪt]   v. 预言；预测
13. decode   [ˌdiːˈkəʊd]   v. 解（码）

Notes 注释

1. central processing unit   中央处理单元
2. arithmetic-logic unit (ALU)   算术逻辑单元
3. internal register   内部寄存器
4. random-access memory   随机存取存储器
5. read-only memory   只读存储器
6. clock driven   时钟驱动
7. digital integrated circuit   数字集成电路
8. binary data   二进制数据

9. input device	输入设备
10. output device	输出设备
11. control unit	控制单元
12. address bus	地址总线
13. data bus	数据总线
14. control bus	控制总线
15. internal register	内部寄存器
16. random-access memory	随机存取存储器
17. read-only memory	只读存储器

Reference Translation 参考译文

1. 定义

微处理器是一种计算机处理器，它将中央处理单元的功能集成在一个或者几个集成电路（IC）上。微处理器是一种多用途、时钟驱动、基于寄存器的数字集成电路。它接收二进制数据作为输入，按照指令对数据进行处理，并提供输出结果。

2. 微处理器的体系结构

微处理器从存储器中读取每一条指令，对其进行解码并加以执行。数据处理是以算术和逻辑运算的形式进行的。依据每条指令，从存储器中取出数据或从输入设备中获取数据，并将处理结果存储在存储器中或传送到适当的输出设备。为了执行所有这些功能，微处理器以适当的方式整合了各种功能单元。这种内部结构称为微处理器的体系结构。

一个典型的微处理器结构如图10-1所示。

1）算术逻辑单元（ALU）：该单元执行加法、减法和诸如"与"和"或"之类的操作。

2）控制单元：该单元从存储器中检索指令代码，并启动ALU执行指令所需的操作序列。

3）总线：通过地址总线，微处理器发送它想要访问的I/O设备或内存位置的地址。微处理器使用数据总线发送和接收来自不同设备（I/O和存储器）的数据和指令。控制总线用于在微处理器和系统中的各种设备之间发送和接收控制信号。

4）内部寄存器：微处理器中通常包含很多寄存器，用于在程序执行期间临时存储数据、指令和地址。

5）输入/输出（I/O）设备：用于数据的输入/输出。输入设备用于将数据输入微处理器，输出设备用于将计算结果传送到外部世界。

6）存储器：存储器用于存储要执行的指令和所涉及的数据。它通常由RAM（随机存取存储器）和ROM（只读存储器）组成。

3. 微处理器的应用

传统上与计算机无关的数千个设备现在都包含了微处理器。图10-2显示了微处理器的一些应用。

树莓派是微处理器的一个有趣的应用，它是一个由树莓派基金会开发的小型单板计算机，以促进学校和发展中国家的计算机基础教学。原型机比最初预期的更受欢迎，甚至被销售到目标市场之外，比如用于机器人等。图10-3展示了一个树莓派。

Unit 10 Microcomputer and Microprocessor
微机和微处理器

Part 2 Single Chip Microcontroller
单片机

1. Definition and Application of Microcontroller

A single chip microcontroller(or microcontroller for short) is a VLSI(Very Large Scale Integration) Integrated Circuit(IC) that contains CPU(Central Processing Unit), Memory, I/O ports and few other components. Figure 10-4 shows an illustration of microcontroller.

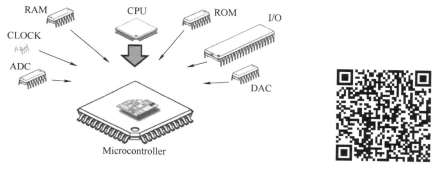

Figure 10-4 Microcontroller

Video 39

Microcontrollers make it economical to digitally control a large number of devices and processes. Microcontrollers are mainly used in automatically controlled products and devices, such as automobile engine control systems, implantable medical devices, remote controls, office machines, appliances, power tools, toys and other embedded systems.

2. Basic Structure of a Microcontroller

Figure 10-5 shows the basic structure of a Microcontroller. Three important components of a microcontroller are CPU, memory and I/O ports.

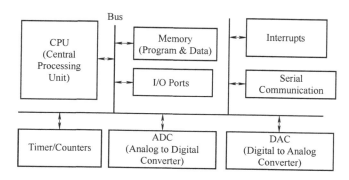

Figure 10-5 Basic Structure of a Microcontroller

(1) CPU   CPU is the brain of the Microcontroller. It consists of an Arithmetic Logic Unit (ALU) and a Control Unit(CU). A CPU reads, decodes and executes instructions to perform

arithmetic, logic and data transfer operations.

(2) Memory Any computational system requires two types of memory: program memory and data memory. Program memory contains the program, i. e. the instructions to be executed by the CPU. Data memory is used to store temporary data while executing the instructions.

(3) I/O ports The interface for the microcontroller to the external world is provided by the I/O ports. Inputs device like switches, keypads, etc. provide information from the user to the CPU in the form of binary data.

The CPU, upon receiving the data from the input devices, executes appropriate instructions and gives response through output devices like LEDs, Displays, Printers, etc.

3. Types of Microcontroller

There are several dozens of microcontroller architectures and vendors, and we will describe two widely-used ones, Intel P8051 microcontroller and ARM microcontroller.

(1) Intel P8051 microcontroller The Intel MCS-51(commonly termed 8051), as shown in Figure 10-6a, is a single chip microcontroller series developed by Intel in 1980 for use in embedded systems. It is an example of a complex instruction set computer, and has separate memory spaces for program instructions and data(Harvard Architecture).

(2) ARM microcontroller Figure 10-6b shows an ARM microcontroller. ARM is a family of reduced instruction set computing(RISC) architectures for computer processors, configured for various environments. With over 100 billion ARM processors produced as of 2017, ARM is the most widely used instruction set architecture in terms of quantity produced.

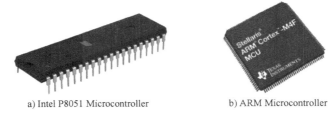

a) Intel P8051 Microcontroller   b) ARM Microcontroller

Figure 10-6 Intel P8051 Microcontroller and ARM Microcontroller

Vocabulary 词汇

1. economical	[ˌiːkəˈnɒmɪkl]	adj.	经济的；合算的
2. implantable	[ɪmˈplɑːntəbl]	adj.	可植入的；可移植的
3. embedded	[ɪmˈbedɪd]	adj.	嵌入的
4. dozen	[ˈdʌzn]	n.	（一）打；十二个
5. vendor	[ˈvendə(r)]	n.	供应商
6. digitally	[ˈdɪdʒɪtəli]	adv.	数字地
7. switch	[swɪtʃ]	n.	（电路的）开关，闸，转换器
8. keypad	[ˈkiːpæd]	n.	（计算机）辅助键盘

Unit 10 Microcomputer and Microprocessor

微机和微处理器

Notes　注释

1. single chip microcontroller　　　　　　单片机
2. very large scale integration（VLSI）　　超大规模集成
3. central processing unit　　　　　　　　中央处理器
4. complex instruction set　　　　　　　　复杂指令集
5. Harvard Architecture　　　　　　　　　哈佛体系结构
6. reduced instruction set computing（RISC）　精简指令集计算
7. I/O port　　　　　　　　　　　　　　　输入/输出端口
8. automobile engine control system　　　 汽车发动机控制系统
9. program memory　　　　　　　　　　　　程序存储器
10. data memory　　　　　　　　　　　　　 数据存储器
11. embedded system　　　　　　　　　　　 嵌入式系统

Reference Translation　参考译文

1. 单片机的定义和应用

单片机（又称微控制器）是一种 VLSI（超大规模集成）集成电路（IC），它包含 CPU（中央处理器）、存储器、I/O 端口以及少量其他组件。图 10-4 所示为单片机示意图。

单片机使数字控制大量的设备和过程变得经济。单片机主要应用于汽车发动机控制系统、植入式医疗设备、遥控器、办公设备、家用电器、电动工具、玩具等嵌入式系统等自动控制产品和设备中。

2. 单片机的基本结构

图 10-5 所示为单片机的基本结构框图。单片机的三个重要组件是 CPU、存储器和 I/O 端口。

（1）CPU（中央处理器）　中央处理器 CPU 是单片机的大脑。它由算术逻辑单元（ALU）和控制单元（CU）组成。CPU 读取、解码和执行指令以执行算术、逻辑和数据传输操作。

（2）存储器　任何计算系统都需要两种类型的存储器：程序存储器和数据存储器。程序存储器包含程序，即 CPU 要执行的指令。数据存储器用来在执行指令时，存储临时数据。

（3）I/O 端口　单片机与外部世界的接口由 I/O 端口或输入/输出端口提供。输入设备，如开关、键盘等，以二进制数据的形式向 CPU 提供用户的信息。

CPU 接收到输入设备的数据后，执行适当的指令，并通过输出设备（如 LED、显示器、打印机等）作出响应。

3. 单片机的类型

目前有几十个单片机结构和供应商，我们将介绍两个广泛使用的单片机：英特尔 P8051 单片机和 ARM 单片机。

（1）Intel P8051 单片机　如图 10-6a 所示，英特尔 MCS-51（通常称为 8051）是英特尔公司于 1980 年开发的用于嵌入式系统的单片机系列。它是一个将复杂指令集成到计算机中的例子，其程序指令和数据有单独的存储空间（哈佛体系结构）。

（2）ARM 单片机　图 10-6b 所示为 ARM 单片机。ARM 是为计算机处理器配置的一系列简化指令集计算（RISC）体系架构，适用于各种环境。截至 2017 年年底，已经生产了超过 1000 亿台 ARM 处理器，从生产数量来看，ARM 是使用最广泛的指令集架构。

Part 3　Programmable Logic Device
可编程逻辑器件

1. Definition

A programmable logic device(PLD) is an electronic component used to build reconfigurable digital circuits. A widely used PLD based on gate array technology is the field-programmable gate array (FPGA).

FPGAs are semiconductor devices that are based around a matrix of configurable logic blocks(CLBs) connected via programmable interconnects. FPGAs can be reprogrammed to desired application or functionality requirements after manufacturing. Figure 10-7 shows an illustration of FPGA.

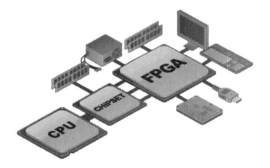

Figure 10-7　An Illustration of FPGA

Video 40

2. Structure

Figure 10-8 shows the FPGA Architecture. The general FPGA architecture consists of three types of modules. They are I/O pads, switch matrix, interconnection wires and configurable logic blocks.

The basic FPGA architecture has two dimensional arrays of CLBs with a means for a user to arrange the interconnection between the logic blocks.

The basic element of the CLB is the Look Up Table(LUT) based function generator, as shown in Figure 10-9. The number of inputs to the LUT vary from 3, 4, 6, and even 8. Now, we have adaptive LUTs that provides two outputs per single LUT with the implementation of two function generators.

3. Major Manufacturers

By 2017, long-time industry rivals Xilinx and Altera were the FPGA market leaders. At that time, they controlled nearly 90 percent of the market.

Unit 10 Microcomputer and Microprocessor
微机和微处理器

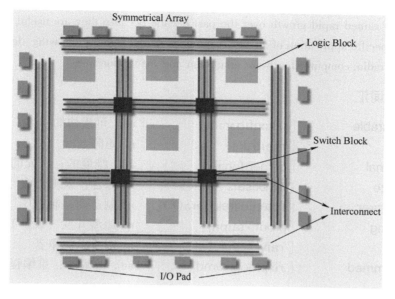

Figure 10-8 FPGA Architecture

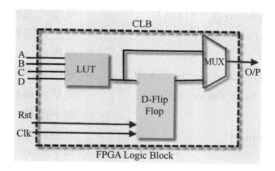

Figure 10-9 FPGA Configurable Logic Block

Both Xilinx and Altera provide proprietary Windows and Linux design software which enables engineers to design, analyze, simulate, and synthesize their designs. Figure 10-10a and Figure 10-10b show FPGA chips manufactured by Altera and Xilinx.

a) Altera FPGA Chip   b) Xilinx FPGA Chip

Figure 10-10 Altera and Xilinx FPGA Chips

4. Application

FPGAs have gained rapid growth over the past decade because they are useful for a wide range of applications. Specific application of an FPGA includes digital signal processing, device controllers, software-defined radio, computer hardware emulation and many more.

Vocabulary　词汇

1. reconfigurable　　　[rɪˈkənfɪɡərəbl]　　　adj. 可重构的
2. matrix　　　　　　[ˈmeɪtrɪks]　　　　　n. 矩阵
3. dimensional　　　　[dɪˈmenʃənəl]　　　　adj. 维度的；次元的
4. synthesize　　　　 [ˈsɪnθəsaɪz]　　　　　v. 综合；人工合成
5. manufacturer　　　 [ˌmænjuˈfæktʃərə(r)]　n. 制造商，制造厂
6. prototyping　　　　[ˈprəʊtəˌtaɪpɪŋ]　　　n. 原型开发（设计）
7. rival　　　　　　　[ˈraɪvl]　　　　　　　n. 对手；竞争者
8. reprogrammed　　　[ˌriːˈprəʊɡræmd]　　　adj. 改编的；重编程的

Notes　注释

1. programmable logic device（PLD）　　　　可编程逻辑器件
2. field-programmable gate array（FPGA）　　现场可编程门阵列
3. configurable logic blocks（CLB）　　　　　可配置逻辑块
4. look up table（LUT）　　　　　　　　　　查找表
5. application specific integrated circuit（ASIC）　专用集成电路
6. simple programmable logic device（SPLD）　简单可编程逻辑器件
7. digital circuit　　　　　　　　　　　　　数字电路
8. programmable interconnect　　　　　　　可编程互连
9. switch matrix　　　　　　　　　　　　　开关矩阵
10. interconnection wire　　　　　　　　　　互连线
11. I/O pad　　　　　　　　　　　　　　　输入/输出焊盘
12. function generator　　　　　　　　　　 函数发生器
13. software-defined radio　　　　　　　　　软件定义无线电
14. computer hardware emulation　　　　　　计算机硬件仿真

Reference Translation　参考译文

1. 定义

可编程逻辑器件（PLD）是用于构建可重建数字电路的电子器件。一种广泛使用的可编程逻辑器件是基于门阵列技术的现场可编程门阵列（FPGA）。

FPGA 是基于通过可编程互连连接的可配置逻辑块（CLB）矩阵的半导体器件。FPGA 可在制造后重新编程为所需的应用或功能要求。图 10-7 所示为 FPGA 示意图。

2. 结构

图 10-8 显示了 FPGA 架构。一般的 FPGA 由三种类型的模块组成。它们是 I/O 焊盘、开

Unit 10　Microcomputer and Microprocessor

微机和微处理器

关矩阵、互连线和可配置逻辑块。

基本的 FPGA 架构具有二维可配置逻辑块阵列，用户可以通过它来安排逻辑块之间的互连。

可配置逻辑块的基本元素是基于查找表（LUT）的函数生成器，如图 10-9 所示。LUT 的输入数量从 3、4、6 甚至到 8 都有。现在，我们有了自适应 LUT，它通过实现两个函数生成器为每个 LUT 提供两个输出。

3. 主要制造商

截至 2017 年年底，长期的行业竞争对手 Xilinx 和 Altera 成为 FPGA 市场的领导者。当时，它们控制了近 90% 的市场。

Xilinx 和 Altera 都提供专有的 Windows 和 Linux 设计软件，使工程师能够设计、分析、模拟和合成它们的设计。图 10-10a、b 所示为 Altera 和 Xilinx 制造的 FPGA 芯片。

4. 应用

FPGA 在过去 10 年中获得了快速增长，因为它们有着广泛的应用。FPGA 的具体应用包括数字信号处理、设备控制器、软件定义无线电和计算机硬件仿真等。

Part 4　Artificial Intelligence(AI) Chip
人工智能芯片

1. Background

Artificial Intelligence(AI) is a scientific technology that studies and develops theories, methods, technologies, and application systems for simulating, extending, and expanding human intelligence.

In recent years, stimulus such as the rise of big data, the innovation of theoretical algorithms, the improvement of computing capabilities, and the evolution of network facilities brought revolutionary progresses to the AI industry which has accumulated knowledge for over 50 years. Research and application innovations surrounding AI has entered a new stage of development.

Video 41

Whether it is the realization of algorithms, the acquisition and a massive database, or the computing capability, the secret behind the rapid development of the AI industry lies in the one and only physical basis, that is, the chips.

2. Definition

At present, there is no strict and widely accepted standard for the definition of AI chips. A broader view is that all chips for AI applications can be called AI chips. Nowadays, some chips based on traditional computing architectures are combined with various hardware and software acceleration schemes, which have achieved great success in some AI application scenarios.

3. Classification

The AI chips mainly include three types:

1) Universal chips that can support AI applications efficiently through hardware and software

optimization, such as GPU(Graphic Process Unit).

2) Chips which focus on accelerating machine learning (especially neural networks and deep learning).

3) Neuromorphic computing chips inspired by biological brain.

4. Examples of AI Chips

(1) Cambricon   Cambricon is the first "neural network" processor chip released by Institute of Computing Technology, Chinese Academy of Sciences, which can "learn in depth" in the world. A Deep Learning processor is a chip that imitates the human brain's multi-layer large-scale artificial neural network. A Cambricon chip is shown in Figure 10-11.

(2) Speech synthesizer chip   XFS5152CE is a new integrated speech synthesizer chip developed by IFLYTEK CO., LTD., which can synthesize Chinese and English speech. It also integrates speech coding and decoding functions to support users to record and play. A XFS5152CE chip is shown in Figure 10-12.

(3) Google TPU   Google Tensor Processing Unit(TPUv1) is a cloud AI chip currently used for all kinds of AI inference in the cloud, such as search queries, translation, and even AlphaGo match.

Figure 10-11   A Cambricon Chip

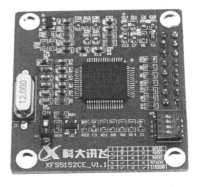

Figure 10-12   A XFS5152CE Chip

Vocabulary   词汇

1. artificial	[ˌɑːtɪˈfɪʃl]	adj.	人造的
2. intelligence	[ɪnˈtelɪdʒəns]	n.	智力；理解力
3. stimulus	[ˈstɪmjələs]	n.	刺激物；刺激因素
4. innovation	[ˌɪnəˈveɪʃn]	n.	改革，创新
5. theoretical	[ˌθɪəˈretɪkl]	adj.	理论的
6. broad	[brɔːd]	adj.	较广阔的，较广泛的
7. strict	[strɪkt]	adj.	精确的；绝对的
8. synthesizer	[ˈsɪnθəsaɪzə(r)]	n.	合成器；综合器
9. scenario	[səˈnɑːriəʊ]	n.	场景

Unit 10　Microcomputer and Microprocessor

Notes　注释

1. GPU（Graphic Process Unit）　　　　　图形处理单元
2. tensor processing unit　　　　　　　　张量处理单元
3. Cambricon　　　　　　　　　　　　　寒武纪（深度神经网络处理器）
4. speech synthesizer chip　　　　　　　语音合成芯片
5. neural network　　　　　　　　　　　神经网络
6. neuromorphic computing chip　　　　神经形态计算芯片
7. search query　　　　　　　　　　　　搜索查询

Reference Translation　参考译文

1. 背景

人工智能（Artificial Intelligence，AI）是研究、开发用于模拟、延伸和扩展人类智能的理论、方法、技术及应用系统的一门科学技术。

近些年，大数据的兴起、理论算法的创新、计算能力的提升及网络设施的演进，使得持续积累了半个多世纪的人工智能产业又一次迎来革命性的进步。人工智能的研究和应用进入全新的发展阶段。

实际上，人工智能产业得以快速发展，无论是算法的实现、海量数据的获取和存储，还是计算能力的体现，都离不开目前唯一的物理基础——芯片。

2. 定义

目前，关于 AI 芯片的定义并没有一个严格和公认的标准。比较宽泛的看法是，面向人工智能应用的芯片都可以称为 AI 芯片。时下，一些基于传统计算架构的芯片和各种软硬件加速方案相结合，在一些 AI 应用场景下取得了巨大成功。

3. 分类

AI 芯片主要包括三类：

1）经过软硬件优化可以高效支持 AI 应用的通用芯片，例如图形处理器 GPU。
2）侧重加速机器学习（尤其是神经网络、深度学习）算法的芯片。
3）受生物大脑启发设计的神经形态计算芯片。

4. AI 芯片举例

（1）寒武纪　寒武纪是中国科学院计算技术研究所发布的世界上第一个"神经网络"处理器芯片，可以"深入学习"。深度学习处理芯片是模拟人脑多层大规模人工神经网络的芯片。寒武纪芯片如图 10-11 所示。

（2）语音合成器芯片　XFS5152CE 是科大讯飞开发的一种新型的综合语音合成器芯片，可以进行中英文语音的合成。它还集成了语音编码和解码功能，支持用户录制和播放。XFS5152CE 芯片如图 10-12 所示。

（3）谷歌 TPU　Google TPU 是面向云端应用的 AI 芯片，可以支持搜索查询、翻译等应用，也是 AlphaGo 的幕后英雄。

Unit 11 Industrial Control Core

工业控制核心

Part 1 Programmable Logic Controller
可编程序控制器

1. Definition

A Programmable Logic Controller(PLC) is an industrial computer control system that continuously monitors the state of input devices and makes decisions based upon a custom program to control the state of output devices. Almost any production line, machine function, or process can be greatly enhanced using this type of control system.

PLCs were first developed in the automobile manufacturing industry to provide flexible and easily programmable controllers to replace hard-wired relays, timers and sequencers. Since then, they have been widely adopted as high-reliability automation controllers suitable for harsh environments.

2. Characteristics

PLC has many advantages suitable for industrial control, as follows.

1) Easy to use, simple modular assembly and connection;

2) Modular expansion capacity of the input, outputs and memory;

3) Simple programming environments and the use of standardized task libraries and debugging aids;

4) Communication capability with other programmable controllers and computers.

Video 42

3. Typical Products

There are different types of PLC. Typical PLC products used in China are listed in Table 11-1.

Table 11-1 Typical PLC Products Used in China

Manufacturer	Product models
OMRON	C series
MTSUBISHI	FX series, Q series
ROCKWELL	MicroLogix1500、SLC-500、CompactLogix、ControlLogix series
GE	90-70 series
SIEMENS	S7 series

4. Structure

The structure of a PLC can be divided into four parts. They are input/output modules, central processing unit(CPU), memory and programming terminal. A programmable controller operates by examining the input signals from a process and carrying out logic instructions on these input signals, producing output signals to drive process equipment. Standard interfaces built-in to PLC allow them to be directly connected to process actuators and sensors without the need for relays. Figure 11-1 shows the structure of the PLC.

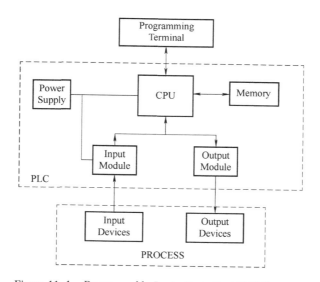

Figure 11-1 Programmable Logic Controller(PLC) Structure

Taking Siemens S7-1500 PLC as an example, S7-1500 PLC hardware system mainly includes power module, CPU module, signal module, communication module, process module and distributed module(such as ET200SP and ET200MP). Up to 32 modules can be installed on the central frame of SIMATIC S7-1500 PLC. The appearance of S7-1500 PLC is shown in Figure 11-2, and the display panel of S7-1500 PLC is shown in Figure 11-3.

5. Operation Sequence

All PLCs have four basic stages of operations that are repeated many times per second, as shown in Figure 11-4.

Figure 11-2　The Appearance of S7-1500 PLC　　Figure 11-3　The Display Panel of S7-1500 PLC

1—LED　2—Display　3—Buttons

1) Self-check: Initially when turned on the first time, a PLC will check its own hardware and software for faults.

2) Input scan: If there are no problems in the self-check stage, a PLC will copy all the input and copy their values into memory, this is called the input scan.

3) Logic scan: Using only the memory copy of the inputs, the ladder logic program will be solved once, this process is called the logic scan.

4) Output scan: While solving the ladder logic, the output values are only changed in temporary memory. When the ladder scan is done, the outputs will be updated using the temporary values in memory, this is called the output scan.

The PLC now restarts the process by starting a self-check for faults. This process typically repeats 10 to 100 times per second.

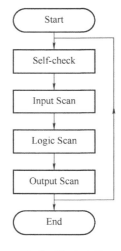

Figure 11-4　Typical Operation Sequence of a PLC

6. Maintenance

Maintenance of programmable controller systems includes only a few basic procedures, which will greatly reduce the failure rate of system components. Preventive maintenance for the PLC system should be scheduled with the regular equipment maintenance. The following are guidelines for preventive measures:

1) Periodically clean or replace any filters that have been installed in enclosures at a frequency dependent on the amount of dust in the area.

2) Do not allow dirt and dust to accumulate on the PLC's components.

3) Periodically check the connections to the I/O modules to ensure that all plugs, sockets, terminal strips, and modules have good connections.

4) Make sure that unnecessary items are kept away from the equipment inside the enclosure.

Unit 11 Industrial Control Core
工业控制核心

Vocabulary 词汇

1. manufacturing [ˌmænjuˈfæktʃərɪŋ] n. 制造业
2. reliability [rɪˌlaɪəˈbɪlɪti] n. 可靠性
3. timer [ˈtaɪmə(r)] n. 计时器；定时器
4. sequencer [ˈsiːkwənsə(r)] n. 序列发生器
5. harsh [hɑːʃ] adj. 严酷的；恶劣的；艰苦的
6. module [ˈmɒdjuːl] n. 模块；功能块；程序块；组件；配件
7. interface [ˈɪntəfeɪs] n. （人机）界面；接口；接口程序；连接电路
8. temporary [ˈtemprəri] adj. 暂时的；临时的
9. guideline [ˈɡaɪdlaɪn] n. 指导原则；准则
10. periodically [ˌpɪəriˈɒdɪkəli] adv. 定期地；周期性地
11. enclosure [ɪnˈkləʊʒə(r)] n. 外壳
12. socket [ˈsɒkɪt] n. 插座；插口
13. security [sɪˈkjʊərəti] n. 保护措施；安全

Notes 注释

1. programmable logic controller (PLC) 可编程序控制器
2. industrial computer control system 工业计算机控制系统
3. assembly line 装配线
4. fault diagnosis 故障诊断
5. hard-wired relay 硬接线继电器
6. ladder logic program 梯形逻辑程序
7. failure rate 故障率
8. preventive maintenance 预防性维护
9. production line 生产线
10. machine function 机器功能
11. automobile manufacturing industry 汽车制造工业
12. automation controller 自动控制器
13. modular assembly 模块化组装
14. modular expansion capacity 模块扩展能力
15. standardized task libraries 标准化任务库
16. input/output module 输入/输出模块
17. programming terminal 编程终端
18. logic instruction 逻辑指令
19. process equipment 过程设备
20. self-check 自检查
21. input scan 输入扫描

22. logic scan　　　　　　　　　　　　　　逻辑扫描
23. output scan　　　　　　　　　　　　　输出扫描

Reference Translation 参考译文

1. 定义

可编程序控制器（PLC）是一种工业计算机控制系统，它连续监测输入设备的状态，并根据自定义程序做出决策，以控制输出设备的状态。使用这种控制系统，几乎任何生产线、机器功能或工艺都可以从中受益。

PLC 最早是由汽车制造工业开发出来的，用于提供灵活且易于编程的控制器，以取代硬接线继电器、定时器和序列器。自那以后，PLC 作为适应恶劣环境的高可靠性自动控制器，被广泛地使用。

2. 主要特点

PLC 具有许多适合工业控制的独特优点，其特点如下。

1）易于使用，模块化组装，连接简单。
2）具有输入、输出和存储器的模块化扩展能力。
3）简单的编程环境以及标准化任务库和调试辅助工具的使用。
4）具有与其他可编程序控制器和计算机的通信能力。

3. 常用 PLC 介绍

PLC 的种类繁多，在我国应用较多的典型 PLC 产品见表 11-1。

表 11-1　典型 PLC 产品

生 产 厂 家	产 品 型 号
日本欧姆龙（OMRON）公司	C 系列
日本三菱（MTSUBISHI）公司	FX 系列，Q 系列
美国罗克韦尔（ROCKWELL）公司	MicroLogix1500、SLC-500、CompactLogix、ControlLogix 系列
美国通用电气（GE）公司	90-70 系列
德国西门子（SIEMENS）公司	S7 系列

4. 结构

PLC 的结构可分为四个部分，它们是输入/输出模块、中央处理器（CPU）、存储器和编程终端。可编程序控制器通过检查来自过程的输入信号并对这些输入信号执行逻辑指令来操作，产生输出信号以驱动过程设备。内置到 PLC 的标准接口允许它们直接连接到过程执行器和传感器，而无需继电器。图 11-1 所示为 PLC 的结构。

以西门子 S7-1500 PLC 为例，S7-1500 PLC 硬件系统主要包括电源模块、CPU 模块、信号模块、通信模块、过程模块和分布式模块（如 ET200SP 和 ET200MP）。SIMATIC S7-1500 PLC 的中央机架上最多可以安装 32 个模块。S7-1500 PLC 的外观如图 11-2 所示，S7-1500 PLC 的显示面板如图 11-3 所示。

5. 操作顺序

所有 PLC 都有 4 个基本操作阶段，每秒重复多次，如图 11-4 所示。

1）自检查：当打开 PLC 时，PLC 会检查自己的硬件和软件是否有故障。

2）输入扫描：如果在自检查阶段没有问题，PLC 将复制所有输入并将其值复制到内存中，这称为输入扫描。

3）逻辑扫描：仅使用输入的存储器副本，梯形图逻辑程序将被处理一次，这称为逻辑扫描。

4）输出扫描：在处理梯形图逻辑时，输出值只在临时内存中更改。当阶梯扫描完成后，输出将使用内存中的临时值进行更新，这称为输出扫描。

4 个阶段完成之后，PLC 通过启动故障自检来重新启动进程。这个过程通常每秒重复 10 ~ 100 次。

6. 维护

可编程序控制器系统的预防性维护只包括一些基本步骤，这将大大降低系统部件的故障率。PLC 系统的预防性维护应安排定期的机器或设备维护。以下是常用维护措施：

1）定期清洁或更换安装在外壳中的过滤器，其频率取决于该区域的灰尘量。

2）不允许污垢和灰尘积聚在 PLC 部件上。

3）定期检查与 I/O 模块的连接，确保所有插头、插座、接线板和模块连接良好。

4）确保不必要的物品远离设备的外壳。

Motion Controller
运动控制器

1. Introduction

Motion Control usually refers to the conversion of scheduled control schemes and planning instructions into desired mechanical motion under complex conditions. The goal of motion control is to achieve precise position control, speed control, acceleration control, torque control of mechanical motion.

Video 43

The programmable motion controller is the brain of the motion system and controls the motor. The motion controller is programmed to accomplish a specific task for a given application. This controller reads a feedback signal to monitor the position of the load. By comparing a desired position with the feedback position, the controller can take action to minimize an error between the actual and desired load positions.

2. Classification

Generally, there are three basic structure schemes of motion control system: motion control based on PLC, motion control based on PC and motion control card, and pure PC control.

(1) Motion control based on PLC There are two kinds of motion control by PLC, as shown in Figure 11-5.

1) The motor is driven by the output impulse of the specific output port of the PLC, and the closed-loop position control of the motor is realized by the high-speed impulse input port.

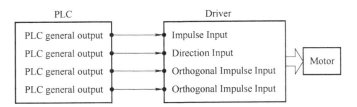

Figure 11-5　Motion Control Based on PLC

2) The closed-loop position control of motor is realized by using the position module extended by PLC.

(2) Motion control based on PC and motion control card　Motion controller is mainly based on motion control card. Industrial PC only provides interpolation operation and motion instructions. Motion control card completes speed control and position control. As shown in Figure 11-6.

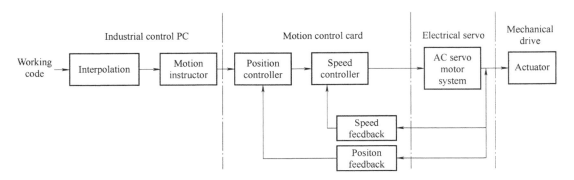

Figure 11-6　Motion Control Based on PC and Motion Control Card

(3) Pure PC control　Figure 11-7 is a fully software-based motion control system using PC. On the high-end hardware platform of industrial PC and embedded PC, the functions of PLC and motion control can be realized by software program, and the logical control and motion control can be realized.

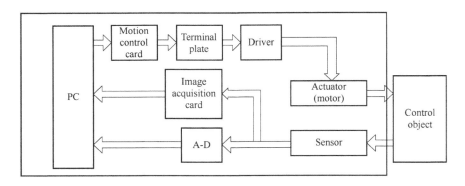

Figure 11-7　Pure PC Motion Control

Unit 11 Industrial Control Core
工业控制核心

Motion controllers receive control commands from the host computer(PC), position information from the position sensor, and output motion commands to the servo motor power drive circuit.

Dedicated motion controller allows most of the calculation to be completed by the chip in the motion controller. They make the hardware design of the control system simple, reduce the data communication between the host computer, and improve the efficiency of the system.

Vocabulary 词汇

1. conversion [kən'vɜːʃn] n. 转变；转换；转化
2. schedule ['ʃedjuːl] n. 工作计划；日程安排 v. 安排；为……安排时间
3. instruction [ɪn'strʌkʃn] n. 指令
4. acceleration [əkˌseləˈreɪʃn] n. 加速；加快；加速度
5. accomplish [əˈkʌmplɪʃ] v. 完成
6. minimize [ˈmɪnɪmaɪz] v. 使减少到最低限度

Notes 注释

1. motion control 运动控制
2. mechanical motion 机械运动
3. mechanical drive 机械传动
4. position control 位置控制
5. speed control 速度控制
6. acceleration control 加速度控制
7. torque control 转矩控制
8. programmable motion controller 可编程运动控制器
9. feedback signal 反馈信号
10. impulse input 脉冲输入
11. direction input 方向输入
12. orthogonal impulse input 正交脉冲输入
13. motion control card 运动控制卡
14. embedded PC 嵌入式计算机
15. industrial PC 工业计算机
16. data communication 数据通信
17. hardware design 硬件设计

Reference Translation 参考译文

1. 介绍

运动控制通常是指在复杂条件下将预定的控制方案、规划指令转变成期望的机械运动。运动控制的目的是实现机械运动的精确位置控制、速度控制、加速度控制、转矩或力的控制。

可编程运动控制器是运动系统的大脑，控制电动机进行运动。运动控制器被编程以完成给定应用程序的特定任务。运动控制器读取反馈信号以监控负载的位置。通过将所需位置与反馈位置进行比较，控制器可以采取措施，将实际和所需负载位置之间的误差降至最低。

2. 分类

一般地，运动控制系统有三种基本结构方案：基于 PLC 的运动控制、基于 PC 和运动控制卡的运动控制、纯 PC 控制。

（1）基于 PLC 的运动控制　PLC 有两种运动控制方式，如图 11-5 所示。

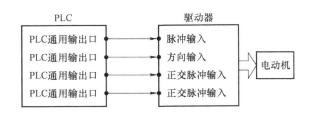

图 11-5　基于 PLC 的运动控制

1）使用 PLC 的特定输出端口输出脉冲驱动电动机，同时使用高速脉冲输入端口来实现电动机的闭环位置控制。

2）使用 PLC 外部扩展的位置模块来进行电动机的闭环位置控制。

（2）基于 PC 和运动控制卡的运动控制　运动控制器以运动控制卡为主，工控 PC 只提供插补运算和运动指令。运动控制卡完成速度控制和位置控制，如图 11-6 所示。

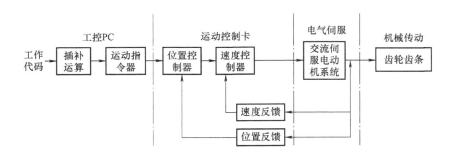

图 11-6　基于 PC 和运动控制卡的运动控制

（3）纯 PC 控制　图 11-7 所示为完全采用 PC 的全软件形式的运动控制系统。在高性能工业 PC 和嵌入式 PC（配备专为工业应用而开发的主板）的硬件平台上，可通过软件程序实现 PLC 和运动控制等功能，实现逻辑控制和运动控制。

运动控制器都从主机（PC）接收控制命令，从位置传感器接收位置信息，向伺服电动机功率驱动电路输出运动命令。

专用运动控制器的使用使得大部分计算工作由运动控制器内的芯片来完成，使控制系统硬件设计简单，与主机之间的数据通信量减少，进而提高了系统效率。

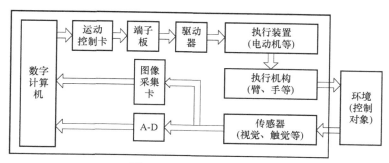

图 11-7 纯 PC 运动控制

Part 3 Human-Machine Interface
人机界面

1. Definition

HMI is the acronym for Human-Machine Interface, an interface between the user and the machine. An HMI is considered an interface; a very broad term that can include MP3 players, industrial computers, household appliances, and office equipment.

However, an HMI is much more specific to manufacturing and process control systems. An HMI provides a visual representation of a control system and provides real time data acquisition. An HMI can increase productivity by having a user-friendly control center.

An HMI is the centralized control unit for manufacturing lines, so that one may access the system at any moment for any purpose. For a manufacturing line to be integrated with an HMI, it must first be working with a PLC. It is the PLC that takes the information from the sensors, so the HMI can make decisions. Illustrations of HMIs are shown in Figure 11-8.

Figure 11-8　HMIs

Video 44

2. Basic Types

There are three basic types of HMIs: the pushbutton replacer, the data handler, and the overseer.

(1) Push-button replacer　Before the HMI came into existence, a control might consist of hundreds of push-buttons and LEDs performing different operations. The pushbutton replacer HMI has streamlined manufacturing processes, centralizing all the functions of each button into one location. An example of push-button replacer is shown in Figure 11-9.

(2) Data handler　Data handler is perfect for applications requiring constant feedback from the system, or printouts of the production reports. With the data handler, you must ensure the HMI screen is big enough for such things as graphs, visual representations and production summaries. The data handler includes such functions as data trending, data logging and alarm handling/logging. An example of data handler is shown in Figure 11-10.

Figure 11-9　Push-Button Replacer　　　　Figure 11-10　Data Handler

(3) Overseer　The application of overseer involve SCADA(Supervisory Control and Data Acquisition) or maybe even MES(manufacturing execution system) functions such as data warehousing, database transactions, and interfacing with an enterprise-type system. With the current emphasis on real-time data analysis and cost reduction, the appropriate HMI could add significant value.

3. Applications

The HMI is used throughout various industries including manufacturing plants, vending machines, food and beverage, pharmaceuticals, and utilities, just to name a few. The integration of the HMI into manufacturing has vastly improved operations.

The HMI allows for supervisory control and data acquisition in the entire system, so parameter changes are feasible as the operator's choosing. For example, in metals manufacturing, an HMI might control how metal is cut and folded, and how fast to do so; An HMI offers improved stock control and replenishment, so fewer journeys are required out to the vendors; HMIs are used in bottling processes to control all aspects of the manufacturing line, such as speed, efficiency, error detection and error correction; Utility companies may use HMIs to monitor water distribution and waste water treatment.

Vocabulary　词汇

1. acronym　　　　　　['ækrənɪm]　　　　　　n. 首字母缩略词

Unit 11 Industrial Control Core
工业控制核心

2. specific	[spəˈsɪfɪk]	adj.	特有的；独特的
3. acquisition	[ˌækwɪˈzɪʃn]	n.	获得，得到
4. streamline	[ˈstriːmlaɪn]	v.	使（系统、机构等）效率更高；（尤指）使增产节约
5. printout	[ˈprɪntaʊt]	n.	（计算机）打印件，打印资料
6. graph	[ɡræf]	n.	图；图表
7. representation	[ˌreprɪzenˈteɪʃn]	n.	表现；描述；描绘
8. warehousing	[ˈweəhaʊzɪŋ]	n.	仓储
9. plant	[plɑːnt]	n.	工厂；（工业用的）大型机器，设备
10. utility	[juːˈtɪləti]	n.	公用事业
11. feasible	[ˈfiːzəbl]	adj.	可行的；行得通的
12. replenishment	[rɪˈplenɪʃmənt]	n.	补充；充满
13. overseer	[ˈəʊvəsɪə(r)]	n.	监督者；监控器

Notes 注释

1. human-machine interface —— 人机界面
2. household appliances —— 家用电器
3. real time data acquisition —— 实时数据采集
4. user-friendly —— 对用户友好的
5. manufacturing line —— 生产线
6. push-button replacer —— 按钮替换器
7. data handler —— 数据处理器
8. data trending —— 数据趋势分析
9. data logging —— 数据记录
10. data warehousing —— 数据仓库
11. database transactions —— 数据库处理
12. real-time data analysis —— 实时数据分析
13. supervisory control and data acquisition —— 监控和数据采集
14. vending machine —— 自动售货机
15. centralized control unit —— 中心控制单元

Reference Translation 参考译文

1. 定义

HMI 是人机界面的缩写，是用户和机器之间的界面。人机界面被认为是一个接口，接口是一个非常广泛的术语，可以包括 MP3 播放器、工业计算机、家用电器和办公设备。

然而，人机界面更是一个制造和过程控制系统。人机界面提供控制系统的可视化表示，并提供实时数据采集。一个 HMI 可以通过一个用户友好的控制中心来提高生产率。

HMI 是生产线的集中控制单元，以便随时出于任何目的访问系统。对于要与 HMI 集成的生产线，必须首先使用可编程序控制器（PLC）。PLC 从传感器中获取信息，以便 HMI 能够做出决定。HMI 的示意图如图 11-8 所示。

2. 基本类型

HMI 有三种基本类型：按钮替换器、数据处理器和监控器。

（1）按钮替换器　在 HMI 出现之前，一个控件可能由数百个按钮和执行不同操作的 LED 灯组成。按钮替换器人机界面简化了生产流程，将每个按钮的所有功能集中到一个位置。一个按钮替换器的示例如图 11-9 所示。

（2）数据处理器　数据处理器非常适合需要系统不断反馈或打印生产报告的应用程序。使用数据处理器时，必须确保 HMI 屏幕足够大，可以进行图形、可视化表示和生产摘要等操作。数据处理器包括数据趋势、数据记录和警报处理/记录等功能。一个数据处理器的例子如图 11-10 所示。

（3）监控器　监控器可实现类似 SCADA（监控和数据采集）甚至 MES（制造执行系统）的功能，如数据仓库、数据库事务和企业系统接口。随着当前对实时数据分析和成本降低的重视，选用合适的人机界面可以产生显著的价值。

3. 应用

HMI 广泛应用于各个行业，包括制造厂、自动售货机、食品和饮料、制药和公用事业等等。人机界面与制造业的集成极大地改善了操作。

HMI 允许在整个系统中进行监控和数据采集，因此可以由操作员通过 HMI 更改参数。例如，在金属制造中，人机界面可以控制金属的切割和折叠方式，以及切割和折叠的速度。HMI 改进了库存控制和补充，因此减少了拜访供应商的次数。HMIS 用于装瓶过程，可以控制生产线的各个方面，如速度、效率、错误检测和错误纠正。公用事业公司可使用 HMI 监测配水和废水处理过程。

Unit 12 Industrial Robot

工业机器人

What are Robots and Robotics?

The word "robot" comes from the play R. U. R. (Rossum's Universal Robots) written by Karel Ĉapek in 1920, which means "forced work or labor." The play began in a factory that used robots. These robots were described as efficient but emotionless, incapable of original thinking and indifferent to self-preservation, as shown in Figure 12-1. This was the earliest idea of industrial robots.

Figure 12-1 A Scene from the Play R. U. R.

Video 45

Today, Robot means any man-made machine that can perform work or other actions normally performed by humans, either automatically or by remote control. Robots are machines that can be used to do jobs. Some robots can do work by themselves. Other robots must always be told what to do.

Robotics is a branch of technologies that deals with the study, design and use of robot systems. Robotics is related to the sciences of electronics, engineering, mechanics, and software. These technologies deal with robots that can take the place of humans in dangerous environments or manufacturing processes, or imitate humans in appearance, behaviour, and cognition. Today, many ideas about robots are inspired by nature, contributing to the field of bio-inspired robotics.

Part 1 About Industrial Robot
关于工业机器人

1. Definition of Industrial Robot

Industrial robot is a robot system used for manufacturing. Industrial robot is defined as "an automatically controlled, reprogrammable, multipurpose, manipulator programmable in three or more axes, which may be either fixed in place or mobile for use in industrial automation applications." in ISO 8373.

The Robotics Institute of America(RIA) defines a robot as: A reprogrammable multi-functional manipulator designed to move materials, parts, tools, or specialized devices through variable programmed motions for the performance of a variety of tasks.

Industrial robot is helpful in material handing and provide interfaces. Typical applications of robots include: welding, painting, assembly, pick and place for PCB, packaging, labeling, palletizing, inspection, and testing; all can be accomplished with high endurance, speed, and precision.

2. History of Industrial Robot

In 1959, George Devol invented Unimate, the first industrial robot, as shown in Figure 12-2.

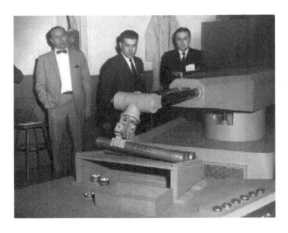

Figure 12-2 The First Industrial Robot, Unimate

In 1961, Unimate robot was used on a production line in the General Motors Corporation in 1961, It was used to lift red-hot door handles and other such car parts from die casting machines and stack them, in an factory in New Jersey, USA.

Unit 12　Industrial Robot
工业机器人

Figure 12-3　PUMA Robot

In 1962, AMF Corporation produced the Verstran robot, which became commercial industrial robots, and exported to countries around the world, setting off a robot boom worldwide.

Around 1970s, many company starts their robotic business and create their first industrial robot. Such as NACHI, KUKA, FANUC, YASKAWA, ASEA(predecessor of ABB), OTC.

In 1978, Unimation creates the PUMA(Programmable Universal Machine for Assembly) robot, as shown in Figure 12-3. PUMA is still working in the production; it marks the maturity of industrial robotics.

The industrial robot industry starts its rapid growth, with a new robot or company entering the market every month.

ABB of Switzerland is one of the world's largest robotics companies. In 1974, ABB developed the world's first fully-electric industrial robot IRB6 and produced the first welding robot in 1975. By 1980, its robotics products were becoming more complete.

Germany's KUKA is one of the world's top manufacturers of industrial robots. In 1973, KUKA developed the first industrial robot, Famulus. KUKA's robots are widely used in the instrument, automotive, aerospace, food, pharmaceutical, medical, casting, plastics and other industries, mainly used in material handling, machine tool equipment, packaging, palletizing, welding, surface finishing and so on.

In 1979, OTC was originally a supplier of welding equipment, they expanded to become a provider to the Japanese auto market of GMAW supplies. Later, OTC Japan introduced its first generation of dedicated arc welding robots. In Japan, 1980 was known as the "first year of robot popularization". Japan began to promote the use of robots in various fields, which greatly alleviated the social contradiction of severe labor shortage in the market. The Japanese government has put forward various incentive policies, and robots are widely used in enterprises.

At present, the most influential and famous industrial robotics companies in the world are mainly divided into European and Japanese. Specifically, they can be divided into "four families" and "four small families", as shown in Table 12-1.

Table 12-1 Examples of Industrial Robot Companies

Category	Company	Country	Logo	Category	Company	Country	Logo
Four families	ABB	Switzerland	ABB	Others	Mitsubishi	Japan	MITSUBISHI ELECTRIC
	KUKA	Germany	KUKA		Epson	Japan	EPSON
	YASKAWA	Japan	YASKAWA		Yamaha	Japan	YAMAHA
	FANUC	Japan	FANUC		Hyundai	Korea	HYUNDAI
Four small families	Panasonic	Japan	Panasonic		CLOOS	Germany	CLOOS
	OTC	Japan	OTC		COMAU	Italy	COMAU
	NACHI	Japan	NACHi		Stäubli	Switzerland	STÄUBLI
	Kawasaki	Japan	Kawasaki		Adept	American	adept

China's industrial robots started in the early 1970s. In the 1980s, Academician Cai Hegao of Harbin University of Technology led the development of China's first arc welding robot- "Huayu I" (HY-I), as shown in Figure 12-4. The key technical problems of trajectory control accuracy and path prediction control were solved in developing the robot. The main technical indicators of the robot reached the international level of similar products.

Figure 12-4 The First Domestic Arc Welding Robot Manufactured by
Harbin University of Technology- Huayu I

Since the early 1990s, with a new round of economic reform and technological progress, industrial robots in China have made a great stride in practice. Industrial robots for various purposes have been developed, a number of robotic application projects have been implemented, and a number of

Unit 12 Industrial Robot
工业机器人

robotic industrialization bases have been built, which laid the foundation of the robotic industry in China.

At present, major companies of industrial robots in China are shown in Table 12-2.

Table 12-2 Major Companies of Industrial Robots in China

Manufacturer	Logo	Company	Logo
SIASUN	SIASUN	GSK	广州数控
ESTUN	ESTUN	HRG	HRG
EFORT	EFORT	BOSHI	BOSHI
ROKAE	ROKAE	PEITIAN	a2
INOVANCE	INOVANCE	AUBO	遨博智能 AUBO

Vocabulary   词汇

1. axes   ['æksɪːz]   n. 轴线；轴心；坐标轴
2. manufacturing   [ˌmænjuˈfæktʃərɪŋ]   n. 制造业；工业 v. 制造；生产
3. automatic   [ˌɔːtəˈmætɪk]   adj. 自动的；无意识的；必然的
4. reprogrammable   [riːˈprəʊɡræməbl]   adj. 可改编程序的；可重复编程的
5. materials   [məˈtɪərɪəlz]   n. 材料；材料科学；材料费
6. interface   [ˈɪntəfeɪs]   n. 接口 v. 使联系
7. typical   [ˈtɪpɪkəl]   adj. 典型的；特有的；象征性的
8. labeling   [ˈleɪblɪŋ]   n. 标签；标记 v. 贴标签；分类
9. inspection   [ɪnˈspekʃn]   n. 视察；检查
10. accomplished   [əˈkʌmplɪʃt]   adj. 完成的；有技巧的
11. distinctive   [dɪˈstɪŋktɪv]   adj. 有特色的，与众不同的
12. poison   [ˈpɔɪzən]   n. 毒药；有毒害的事物
13. commercial   [kəˈmɜːʃəl]   adj. 商业的；营利的 n. 商业广告
14. predecessor   [ˈpriːdɪsesə(r)]   n. 前任，前辈；原先的动画
15. red-hot   [redˈhɒt]   adj. 炽热的；最新的
16. gripper   [ˈɡrɪpə]   n. 夹子；钳子；抓器，抓爪

Notes   注释

1. robot system   机器人系统
2. manipulator programmable   可编程机械手
3. the Robotics institute of America 美国机器人研究所
4. production line   生产线
5. die casting machines   压铸机

6. verstran robot verstran 机器人
7. automatically controlled 自动控制
8. industrial automation applications 工业自动化应用
9. robotics institute 机器人协会
10. materials handling robot 材料处理机器人
11. assembly line 装配线
12. arc welding robots 电弧焊机器人

Reference Translation 参考译文

什么是机器人和机器人学？

"机器人"一词来自卡雷尔·乔佩克1920年写的剧本R. U. R.（罗森的通用机器人），意思是"强迫工作或劳动"。这个剧本开始于一家使用机器人的工厂。这些机器人被描述为高效但无感情，不能进行原始思维，对自我保护漠不关心，如图12-1所示。这是工业机器人最早的想法。

今天，机器人指的是任何人造机器，它可以自动或遥控地执行人类通常执行的工作或其他动作。机器人是可以用来工作的机器。有些机器人可以自己工作。其他机器人必须总是被告知该做什么。

机器人学是研究、设计和使用机器人系统的技术分支。机器人学涉及电子、工程、机械和软件科学。这些技术所涉及的机器人可以在危险环境或制造过程中代替人类，或者在外表、行为和认知上模仿人类。今天，许多关于机器人的想法都是受大自然的启发，为生物启发机器人领域做出了贡献。

1. 工业机器人的定义

工业机器人是一种用于生产制造的机器人系统。工业用机器人在ISO8373中被定义为："一个自动控制、可重复编程的、多用途的、三个或更多个轴的可编程机械手，可以在工业自动化应用中固定或移动使用。"

美国机器人协会（RIA）将机器人定义为："一种用于移动材料、零件、工具或专用设备的，通过可编程序动作来执行各种任务的，具有编程能力的多功能机械手。"

工业机器人有助于物料搬运和提供接口。机器人的典型应用包括焊接、喷涂、组装、PCB板取放、包装、贴标、码垛和检测；机器人都能够以很高的耐久性、高速性和高精确实现这些工作。

2. 工业机器人发展史

1959年，乔治·德沃尔（George Devol）发明了第一个工业机器人Unimate，如图12-2所示。

1961年，Unimate机器人被应用在通用汽车公司的一条生产线上。在美国新泽西州的一家工厂中，它被用来从压铸机中提取红热的门把手和其他类似的汽车零件，并将其堆放起来。

1962年，AMF公司生产了Verstran机器人，该机器人作为商业工业机器人，被出口到世界各国，在全球范围内掀起了机器人热潮。

20世纪70年代，许多公司开展了机器人业务，并制造了它们的第一台工业机器人，如

NACHI、KUKA、FANUC、YASKAWA、ASEA（ABB 的前身）和 OTC。

1978 年，Unimation 公司推出 PUMA（用于装配的可编程通用机器）机器人，如图 12-3 所示。如今，PUMA 仍然工作在生产第一线，这标志着工业机器人技术已经完全成熟。

工业机器人行业开始快速增长，每个月都会有新机器人或公司面世。

瑞士的 ABB 公司是世界上最大的机器人制造公司之一。1974 年 ABB 公司研发了世界上第一台全电控式工业机器人 IRB6，并于 1975 年生产出第一台焊接机器人。到 1980 年，其机器人产品趋于完备。

德国的 KUKA 公司是世界上几家顶级工业机器人制造商之一。1973 年 KUKA 开发了第一台工业机器人——Famulus。所生产的机器人广泛应用在仪表、汽车、航空航天、食品、制药、医疗、铸造和塑料等工业，主要用于材料处理、机床装备、包装、堆垛、焊接和表面修整等。

1979 年，OTC 原本是一家焊接设备供应商，后来扩大成为日本汽车市场 GMAW 的供应商，之后日本 OTC 公司推出了第一代专用弧焊机器人。1980 年被称为日本的"机器人普及元年"，日本开始在各个领域推广使用机器人，大大缓解了劳动力市场严重短缺的社会矛盾，再加上日本政府的多方面鼓励政策，机器人受到了广大企业的欢迎。

目前，国际上较有影响力的、著名的工业机器人公司主要分为欧系和日系两种，具体来说，可分成"四大家族"和"四小家族"两个阵营，见表 12-1。

表 12-1 工业机器人阵营

阵营	企业	国家	标识	阵营	企业	国家	标识
四大家族	ABB	瑞士	ABB	其他	三菱	日本	MITSUBISHI ELECTRIC
	库卡	德国	KUKA		爱普生	日本	EPSON
	安川	日本	YASKAWA		雅马哈	日本	YAMAHA
	发那科	日本	FANUC		现代	韩国	HYUNDAI
四小家族	松下	日本	Panasonic		克鲁斯	德国	CLOOS
	欧地希	日本	OTC		柯马	意大利	COMAU
	那智不二越	日本	NACHi		史陶比尔	瑞士	STÄUBLI
	川崎	日本	Kawasaki		爱德普	美国	adept

我国工业机器人起步于 20 世纪 70 年代初期，80 年代哈工大蔡鹤皋院士主持研制出了我国第一台弧焊机器人——"华宇Ⅰ型"（HY-Ⅰ型），如图 12-4 所示，解决了机器人轨迹控制精度及路径预测控制等关键技术，该机器人的主要技术指标达到了国际同类产品水平。

从 20 世纪 90 年代初期起，伴随新一轮的经济体制改革和技术进步，我国工业机器人在实践中取得了长足的进步，先后研制出了各种用途的工业机器人，并实施了一批机器人应用工程，形成了一批机器人产业化基地，为我国机器人产业的腾飞奠定了基础。

目前，国内主要工业机器人厂商见表12-2。

表12-2 国内主要工业机器人厂商

企业	标识	企业	标识
沈阳新松	SIASUN	广州数控	广州数控 GSK
芜湖埃夫特	EFORT	哈工大机器人集团	HRG
南京埃斯顿	ESTUN	哈尔滨博实	BOSHI
北京珞石	ROKAE	安徽配天	a²
深圳汇川	INOVANCE	北京遨博	遨博智能 AUBO

Part 2 Component and Types of Industrial Robot
工业机器人的组成和分类

1. Component of Industrial Robot

A typical industrial robot consists of a robotic arm, a control system, a teach pendant, an end effector as well as some other peripheral devices, as shown in Figure 12-5.

Video 46

Figure 12-5　Components of Industrial Robot
1—Control system　2—Teach pendant　3—End effector　4—Robotic arm

1) Robotic arm is also known as the manipulator, a type of mechanical arm, usually programmable, with similar functions to a human arm. The arm may be the sum total of the mechanism or may be part of a more complex robot. Robotic arm is connected by joints, which allow either rotational motion or translational(linear) displacement. The terminus of the manipulator is end effector. The robotic arm, basically, is the part that moves the tool, but not every industrial robotic resembles an arm, there

are different types of different robot structures.

2) The control system resembles robot's brains, is also known as the "brain" which is run by a computer program. Usually, the program is very detailed as it gives commands for the moving parts of the robot to follow.

3) Teach pendant is also known as the teaching box or demonstrator, makes up the user environment. The teach pendant is usually used only in time of programming. a human-computer interaction interface, connected with the controller, can be operated to move by the operator.

4) The end effector is also known as the robotic hand, or end-of-arm-tooling(EoAT), is a device designed for a specific task, such as welding, gripping, spinning etc., depending on the application. End effectors are generally highly complex, made to match the handled product and often capable of picking up an array of products at one time. They may utilize various sensors to aid the robot system in locating, handling, and positioning products.

2. Types of Industrial Robot

There are many types of industrial robots, the main types of the most commonly used industrial robots are described as below.

(1) Cartesian Robot/Gantry Robot Cartesian coordinate robot(also called linear robot) is an industrial robot with simple structure, as shown in Figure 12-6. Cartesian coordinate robot is a robot whose arm has three prismatic joints, its axes are coincident with a Cartesian coordinator; its three principal axis of control are linear(i. e. they move in a straight line rather than rotate) and are at right angles to each other. The three sliding joints correspond to moving the wrist up-down, in-out, back-forth. This mechanical arrangement simplifies the robot control solution.

A popular application for cartesian coordinate robot is a computer numerical control machine (CNC machine) and 3D printing, as shown in Figure 12-7.

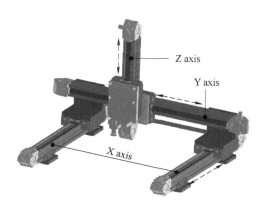

Figure 12-6 Cartesian Coordinate Robot

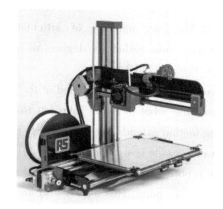

Figure 12-7 3D Printing

(2) SCARA robot The SCARA acronym stands for selective compliance assembly robot arm or selective compliance articulated robot arm, as shown in Figure 12-8. Its arm is slightly compliant in the XY-axes but rigid in the Z-axis, this feature is particularly suitable for assembly work, Therefore,

the SCARA system is widely used for assembling printed circuit boards and electronic components. The other attribute of the SCARA is the jointed two-link arm layout, hence it often uses the term "articulated". This feature allows the arm to extend into confined areas and then retract or "fold up" out of the way.

SCARA trobots are generally faster and cleaner than comparable Cartesian robot systems. Their single pedestal mount requires a small footprint and provides an easy, unhindered form of mounting.

There also exists a so-called double-arm SCARA robot architecture, in which two of the motors are fixed at the base. The first such robot was commercialized by Mitsubishi Electric. In China, HRG company developed five-bar robot HRG-HD1B5B, as shown in Figure 12-9.

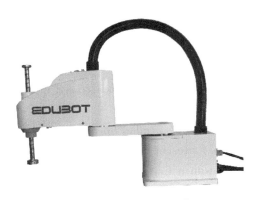
Figure 12-8 SCARA Robot

Figure 12-9 five-bar Robot HRG-HD1B5B

(3) Articulated Robot   An articulated robot is a robot with rotary joints. Articulated robots can range from simple two-jointed structures to systems with 10 or more interacting joints. The six-axis articulated robot is a kind of widely used mechanical equipment. The vast majority of articulated robots have six axes, also called six degrees of freedom, as shown in Figure 12-10.

Axis 1——This axis is located at the robot base and allows the robot to rotate from left to right. This sweeping motion extends the work area to include the area on either side and behind the arm. This axis allows the robot to spin up to a full 180 degree range from the center point. This axis is also known as J1.

Axis 2——This axis allows the lower arm of the robot to extend forward and backward. It is the axis powering the movement of the entire lower arm. This axis is also known as J2.

Axis 3——The axis extends the robot's vertical

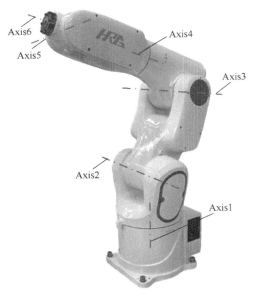
Figure 12-10 Six Axes of Six-Axis Articulated Robot

reach. It allows the upper arm to raise and descend. In some articulated models, it allows the upper arm to reach behind the body, further expanding the work scope. This axis gives the upper arm the better part access. This axis is also known as J3.

Axis 4——This axis works in conjunction with axis 5; it aids in the positioning of the end effector and manipulation of the part. Known as the wrist roll, it rotates the upper arm in a circular motion moving parts between horizontal to vertical orientations. This axis is also known as J4.

Axis 5——This axis allows the wrist of the robot arm to tilt up and down. This axis is responsible for the pitch and yaw motion. The pitch, or bend, motion is up and down, much like opening and closing a box lid. Yaw moves left and right, like a door with hinges. This axis is also known as J5.

Axis 6——This is the wrist of the robot arm. It is responsible for a twisting motion, allowing it to rotate freely in a circular motion, both to position end effectors and to manipulate parts. It is usually capable of more than a 360 degree rotation in either a clockwise or counter clockwise direction. This axis is also known as J6.

Six-axis robots allow for greater flexibility and can perform a wider variety of applications than robots with fewer axes.

(4) Parallel Robot(e. g. delta robot)   The Delta robot is a parallel robot. It consists of multiple kinematic chains connecting the base with the end effector. The robot can also be seen as a spatial generalization of a four-bar linkage. The key concept of the Delta robot is the use of parallelograms which restrict the movement of the end platform to pure translation, i. e. only movement in the X, Y or Z direction with no rotation.

Figure 12-11   ABB-FlexPicker   Figure 12-12   FANUC M-3iA Robot

The robot base is mounted above the workspace and all the actuators are located on it. From the base, three middle jointed arms extend. The end of these arms is connected to a small triangular platform. Actuation of the input links will move the triangular platform along the X, Y or Z direction. Actuation can be done with linear or rotational actuators, with or without reductions (direct drive).

Since the actuators are all located on the base, the arms can be made of a light composite material. As a result of this, the moving parts of the Delta robot have a small inertia. This allows for very high speed and high accelerations.

Delta robots have popular usage in picking and packaging in factories because they can be quite fast. Industries that take advantage of the high speed of Delta robots are the packaging industry, medical and pharmaceutical industry.

Vocabulary 词汇

1. structure	['strʌktʃə]	n.	构造；建筑物
2. selective	[sɪ'lektɪv]	adj.	选择性的；讲究的
3. attribute	[ə'trɪbjuːt]	n.	属性
4. feature	['fiːtʃə(r)]	n.	产品特点，特征
		v.	是……的特色，使突出
5. mount	[maʊnt]	v.	增加；安装，架置 n. 山峰；底座
6. footprint	['fʊtprɪnt]	n.	足迹；脚印
7. end-user	['endˈjuːzə]	n.	终端用户
8. rigid	['rɪdʒɪd]	adj.	僵硬的；坚硬的；精确的
9. interacting	[ˌɪntər'æktɪŋ]	n.	相互作用互相影响
10. vertical	['vɜːtɪkl]	adj.	垂直的；[解剖] 头顶的 n. 垂直面
11. tilt	[tɪlt]	v.	倾斜；翘起 n. 倾斜
12. position	[pə'zɪʃn]	n.	地方；位置 v. 定位；放置
13. manipulation	[məˌnɪpjuˈleɪʃn]	n.	操纵；操作；处理；篡改
14. commercialized	[kəˈmɜːʃəlaɪzd]	adj.	商业化的 v. 使商品化
15. pitch	[pɪtʃ]	v.	倾斜；投掷
16. restrict	[rɪ'strɪkt]	v.	限制；约束；限定
17. rotation	[rəʊ'teɪʃn]	n.	旋转；循环，轮流
18. triangular	[traɪˈæŋɡjələ(r)]	adj.	三角的，[数] 三角形的；三人间的
19. reduction	[rɪ'dʌkʃn]	n.	减少；下降；还原反应
20. inertia	[ɪ'nɜːʃə]	n.	[力] 惯性；迟钝；不活动

Notes 注释

1. pick and place　　　　　　　　　　拾取和放置
2. arc welding　　　　　　　　　　　电弧焊
3. articulated robot　　　　　　　　　铰接式机器人，关节机器人
4. SCARA robots　　　　　　　　　　SCARA 机器人
5. Delta robots　　　　　　　　　　　Delta 机器人
6. cartesian coordinate robots　　　　　笛卡儿坐标机器人
7. gantry robots　　　　　　　　　　龙门式机器人
8. control system　　　　　　　　　　控制系统
9. teach pendant　　　　　　　　　　示教盒

10.	end effector	末端执行器
11.	user environment	用户环境
12.	prismatic joints	棱柱形接头
13.	gas welding	气焊
14.	parallel robot	并联机器人
15.	linear robot	线性机器人
16.	sliding joint	滑动接头
17.	computer numerical control machine	计算机数控机床
18.	double-arm SCARA robot	双臂 SCARA 机器人
19.	assembly robot	装配机器人
20.	two-link arm	联合双链臂
21.	robot base	机器人底座
22.	lower arm	下臂
23.	upper arm	上臂
24.	six degrees of freedom	六自由度
25.	articulated model	关节模型
26.	dual-arm	双臂
27.	yaw motion	偏航运动
28.	bend motion	弯曲运动
29.	four-bar linkage	四杆联动
30.	composite material	复合材料
31.	medical and pharmaceutical industry	医药行业

Reference Translation　参考译文

1. 工业机器人的组成

典型的工业机器人由机器臂、控制系统、示教器、末端执行器以及一些其他外围设备组成，如图 12-5 所示。

1）机械臂，即操作机，是一种机械手臂，通常是可编程的，具有与人类手臂相似的功能。机械臂可以是机械装置的总和，也可以是更复杂机器人的一部分。机械臂通过关节连接，允许产生旋转运动或平移（线性）位移。机械臂的末端是末端执行器。机器人手臂基本上是移动工具的部件，但并不是每个工业机器人都类似于人类的手臂，而是有不同类型的不同机器人结构。

2）控制系统类似于机器人的大脑，也称为由计算机程序运行的"大脑"。通常，程序非常详细，因为它为机器人的运动部件提供了命令。

3）示教器又称为示教盒，构成了用户环境，示教器通常仅在编程时使用。它是与控制器相连的人机交互界面，可由操作员手持操作。

4）末端执行器，即机器手或手臂末端工具（EoAT），是为特定任务而设计的装置，如焊接、夹持、旋转等，具体取决于应用。末端执行器通常非常复杂，需要与要处理的产品相匹配，并且通常能够同时拾取一系列产品。它们可以利用各种传感器来协助机器人系统进行寻迹、搬运和定位产品。

2. 工业机器人的种类

工业机器人有很多种构造，这里介绍最常用的工业机器人构型：

（1）直角坐标机器人（笛卡儿机器人）/龙门机器人　直角坐标机器人（也称为线性机器人）是一种结构简单的工业机器人，如图12-6所示。

笛卡儿坐标机器人是一种机器人，这种机器人的手臂有三个棱柱形关节，轴线与笛卡儿坐标系一致，三个主要控制轴是线性的（即它们以直线而不是旋转的方式移动）并且两两垂直。三个滑动关节对应于手腕上下、进出、前后运动。这种机械装置简化了机器人解决方案。

直角坐标机器人的一个普遍应用是计算机数控机床（CNC机床）和3D打印，如图12-7所示。

（2）SCARA机器人　SCARA是选择顺应性装配机器人手臂或者选择顺应性关节机器人手臂的缩写，如图12-8所示。

SCARA机器人手臂在X，Y轴方向上具有顺应性，而在Z轴方向具有良好的刚性，这种特性特别适合装配工作。SCARA机器人大量应用于装配印制电路板和电子零部件。SCARA机器人的第二个属性是联合双链臂布局（"铰接"）。这个功能允许手臂延伸到狭窄的区域，然后退缩或"折叠"到一边。

SCARA机器人通常比同类笛卡儿机器人系统更快，更洁净。其单基座安装需要一个较小的占地面积，并提供了一种简单、无阻碍的安装形式。

大多数SCARA机器人基于串行架构，这意味着第一台电动机应该携带所有其他电动机。还存在一种所谓的双臂SCARA机器人结构，其中两个电动机固定在基座上。第一台这样的机器人是由三菱电机进行商业化推广的。在国内，哈工大机器人集团推出了HRG-HD1B5B型五杆机器人，如图12-9所示。

（3）铰接式机器人（关节机器人）　关节机器人是一种具有旋转关节的机器人。关节机器人可以从简单的双联结构到具有10个或更多交互关节的系统。六轴关节机器人是一类广泛使用的机械设备。绝大多数关节机器人具有6个轴，也称为六自由度。

轴1—该轴位于机器人底座上，允许机器人从左到右旋转。这种运动把延伸工作区包括在任一侧上以及臂后面的区域。该轴允许机器人从中心点旋转达180°的范围。该轴也称为J1。

轴2—该轴允许机器人的下臂向前和向后延伸。它是为整个下臂运动提供动力的轴。该轴也称为J2。

轴3—该轴延伸了机器人的垂直范围。它允许上臂升高和降低。在一些关节模型中，它允许上臂到达身体后面，进一步扩大了工作范围。该轴使上臂更好地进入部件。该轴也称为J3。

轴4—该轴配合轴5一起工作，它有助于末端执行器的定位和对部件的操纵。它又被称为腕带。它使上臂在水平方向和垂直方向之间以圆周运动的部分旋转。该轴也称为J4。

轴5—该轴允许机器人手臂的手腕向上或向下倾斜。该轴负责俯仰和偏航运动。俯仰或弯曲是上下运动，就像打开或关闭盒盖一样。偏航向左或向右移动，就像有铰链的门。该轴也称为 J5。

轴 6——这是机器人手臂的手腕。它负责扭转运动，允许其以圆周运动自由旋转，既可以定位末端执行器，也可以操纵零件。它通常能够在顺时针或逆时针方向上旋转超过 360°。该轴也称为 J6。

相比轴数较少的机器人，六轴机器人具有更大的灵活性，并且具有更广泛的应用。

（4）并联机器人　Delta 机器人是一种并联机器人，即它由连接底座和末端执行器的多个运动链组成。这种机器人也可以看作四杆机构的空间推广。Delta 机器人的关键概念是使用平行四边形，其将末端平台的移动限制到纯平移，即在没有旋转的情况下仅在 X，Y 或 Z 方向上移动。

机器人底座安装在工作区上方，所有执行器都位于其上。三个中间连接臂从基座开始延伸。这些手臂的端部连接到一个小的三角形平台上。驱动输入连杆将沿着 X，Y 或 Z 方向移动三角形平台。可以通过带或不带减速（直接驱动）的线性或旋转执行器进行驱动。

由于执行器都位于基座上，臂可以由轻质复合材料制成，因此，Delta 机器人的运动部件具有较小的惯性。这允许产生非常高的速度和高加速度。

Delta 机器人在工厂中普遍用于抓取和包装，因为它们可以相当快。利用 Delta 机器人高速运转特性的行业是包装行业、医疗和制药行业。

Part 3　Technical Parameters of Industrial Robot
工业机器人的技术参数

The technical parameters reflect the scope and performance of the robot, including number of axes, degrees of freedom, payload, working envelope, maximum speed, resolution, accuracy, repeatability. other parameters are control mode, drive mode, installation, power source capacity, body weight and environmental parameters.

1. Number of Axes

Two axes are required to reach any point in a plane; three axes are required to reach any point in space. To fully control the orientation of the end of the arm(i. e. the wrist) three more axes are required. Some designs(e. g. the SCARA robot) trade limitations in motion possibilities for cost, speed, and accuracy.

2. Degrees of Freedom(DOF)

It is the independent motions that the end effector of the robot can move. It is defined by the number of axes of motion of the manipulator, usually the same as the number of axes, as shown in Figure 12-13. DOF reflects the flexibility of robot action, with the more DOF, the robot is closer to human action, the versatility of the robot is better. However, the mechanisms are more complex and the overall requirements of the robot are higher. Therefore, the degrees of freedom of industrial robots is designed according to application.

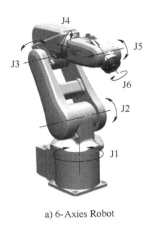
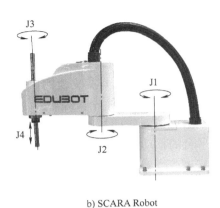

a) 6-Axies Robot b) SCARA Robot

Figure 12-13 DOF of Robot

Video 47

3. Payload

It is the weight that a robot can lift. The maximum payload is the weight carried by the robot manipulator at reduced speed while maintaining rated precision. Nominal payload is measured at maximum speed while maintaining rated precision. The payload of different industrial robots is described in Table 12-3.

Table 12-3 Payload of Industrial Robot

Product number	Payload	Image	Product number	Payload	Image
ABB YuMi	0.5kg		YASKAMA MH12	12kg	
ABB IRB120	3kg		YASKAWA MC2000II	50kg	
KUKA KR6	6kg		FANUC R-2000iB/210F	210kg	

(续)

Product number	Payload	Image	Product number	Payload	Image
KUKA KR16	16kg		FANUC M-200iA/2300	2300kg	

4. Working Envelope

It is also known as working space (reach envelope). A three-dimensional shape that defines the region of space a robot can reach, as shown in Figure 12-14.

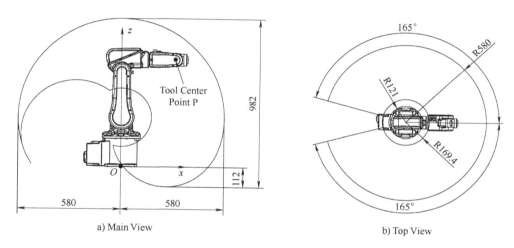

a) Main View　　　　　　　　　　　　b) Top View

Figure 12-14　Working Envelope of ABB-IRB120

5. Maximum Speed

It is the compounded maximum speed of the tip of a robot moving at full extension with all joints moving simultaneously in complimentary directions. This speed is the theoretical maximum and should not be used to estimate cycle time for a particular application.

6. Resolution

It is the smallest increment of motion or distance that can be detected or controlled by the control system of a mechanism. The resolution of any joint is a function of encoder pulses per revolution and drive ratio, and dependent on the distance between the tool center point and the joint axis.

7. Accuracy

It is the closeness a robot can reach a commanded position. Accuracy is the difference between the point that a robot is trying to achieve and the actual resultant position. Absolute accuracy is the difference between a point instructed by the robot control system and the point actually achieved by the manipulator arm, while repeatability is the cycle-to-cycle variation of the manipulator arm when aimed at the same point.

8. Repeatability

It is the degree that the robot will return to a programmed position. That is to say, the ability of a system to repeat the same motion or achieve the same points when presented with the same control signals. It is the cycle-to-cycle error of a system when trying to perform a specific task.

Vocabulary 词汇

1. orientation [ˌɔːriən'teɪʃn] n. 方向；定向；情况介绍
2. flexibility [ˌfleksə'bɪləti] n. 灵活性；弹性；适应性
3. versatility [ˌvɜːsə'tɪləti] n. 多功能性；用途广泛
4. scope [skəʊp] n. 范围；余地；视野
5. payload ['peɪləʊd] n. 有效载荷
6. resolution [ˌrezə'luːʃn] n. 分辨率
7. repeatability [rɪˌpiːtə'bɪlɪti] n. 重复性；再现性
8. installation [ˌɪnstə'leɪʃn] n. 安装；装置；就职仪式
9. increment ['ɪŋkrəmənt] n. 增量；增加；增额

Notes 注释

1. technical parameters — 技术参数
2. degrees of freedom — 自由度
3. maximum payload — 最大有效载荷
4. nominal payload — 额定负载
5. working envelope — 工作空间
6. maximum speed — 最大速度
7. environmental parameters — 环境参数
8. drive ratio — 驱动比
9. absolute accuracy — 绝对精度

Reference Translation 参考译文

技术参数反映了机器人的范围和性能，包括轴数、自由度、有效载荷、工作空间、最大速度、分辨率、精度和重复性等。其他参数有控制方式、驱动方式、安装方式、电源容量、重量和环境参数。

1. 轴数

需要两个轴才能到达平面上的任意点；需要三个轴才能到达空间中的任意点。为了完全控制手臂末端（即手腕）的方向，需要三个以上的轴。有些设计（如 SCARA 机器人）在成本、速度和精度方面限制了运动的可能性。

2. 自由度（DOF）

机器人末端执行器可以移动的独立运动的数量，由机械手的运动轴数定义，通常与轴数相同，如图12-13所示。机器人的自由度反映了机器人动作的灵活性，自由度越多，机器人就越能接近人手的动作机能,通用性越好;但是自由度越多,结构就越复杂,对机器人的整体要求就越高。因此,工业机器人的自由度是根据其用途设计的。

3. 有效载荷

有效载荷是指机器人能举起多少重量。最大有效载荷是在保持额定精度的情况下,机器人机械手速度降低后能承载的重量。额定负载是在最大速度并保持额定精度下进行衡量的。

4. 工作空间

工作空调定义了机器人能够达到的空间区域的三维图形,如图 12-14 所示。

5. 最大速度

所有关节在互补方向上同时移动时,机器人末端在完全伸展情况下的最大复合速度。这个速度是理论上的最大值,不能用于估计特定应用的周期时间。

6. 分辨率

分辨率是指可由机构的控制系统检测或控制的运动或距离的最小增量。任何关节的分辨率是编码器脉冲每转和驱动比的函数,取决于刀具中心点和关节轴之间的距离。

7. 精度

精度是机器人能够到达一个指令位置的距离,是机器人试图达到的与实际达到的位置之间的差异。绝对精度是机器人控制系统目标点与由机械臂实际到达点之间的差异,而重复性是机械手重复到达同一点时的周期变化。

8. 重复性

重复性是指机器人返回编程位置的程度。也就是说,系统重复相同动作或施加同一控制信号时到达相同位置的能力。当尝试执行特定任务时,系统循环错误。

Unit 13 Industry Application of Robot

机器人行业应用

Part 1 Handling Robot
搬运机器人

Handling robots are industrial robots that can be automated for handling operations. The earliest handling robots appeared in 1960 in the United States, the two styles of Versatran and Unimate robots were used for handling operations for the first time. Handling operation is that uses a device to hold the workpiece, and moves it from one process place to another. The handling robot can be fitted with different end actuators to perform various workpieces handling in different shapes and conditions, that

Video 48

greatly reduces the heavy manual labor of human. At present, more than 100,000 handling robots are used in automatic assembly line, palletizing, container and so on. Some developed countries have drawn up the maximum amount of manual handling, the work which is more than the limit must be carried by the handling robot.

The handling robot is a high-tech in the field of modern automatic control, which involves the academic areas of Mechanics, Mechanology, hydraulic-air pressure technology, automatic control technology, sensor technology, single chip microcomputer technology and computer technology, and it has become an important component of modern machinery manufacturing system. Its advantage is that you can program to complete a variety of tasks; they have advantages in their own structures and performances which people and the machine own. In particular, they reflect the artificial intelligence and adaptability.

Commonly, there are tandem joint robots, horizontal joint robots(SCARA robots), Delta parallel

Unit 13 Industry Application of Robot
机器人行业应用

joint robots and AGV handling robots. Above all kinds of robots in the handling industry have their own characteristics.

The tandem joint robot usually has four or six rotating axes, and is similar to the human arm, more flexible, and the load of them is larger; they are used for handling heavier items.

SCARA robot is a horizontal multi-joint robot, and has faster work frequency, that is particularly suitable for handling goods in laboratory automation, medicine, consumer electronics, food, automotive, PC peripherals, semiconductors, plastics, home appliances and aerospace industry.

Delta robot belongs to parallel robot. Delta robots are mainly used for handling, assembling food, medicine and electronic products and so on. Delta robot is widely used in the market because of its light weight, small size, fast moving speed, accurate positioning, low cost and high efficiency.

China is in the critical moment of the industrial transformation and upgrading, more and more enterprises introduce industrial robots in the manufacturing process, it will profoundly affect all aspects of Chinese-made.

Vocabulary 词汇

1. container	[kənˈteɪnə]	n.	集装箱；容器
2. adaptability	[əˌdæptəˈbɪləti]	n.	适应性；可变性；适合性
3. horizontal	[ˌhɒrɪˈzɒntl]	adj.	水平的；地平线的；同一阶层的
		n.	水平线；水平面
4. parallel	[ˈpærəlel]	n.	平行线 v. 使……与……平行
		adj.	平行的；类似的
5. rotate	[rəʊˈteɪt]	v.	旋转；循环
6. plastic	[ˈplæstɪk]	n.	塑料；整形外科；外科修补术
7. guide	[gaɪd]	v.	引导；带领；操纵
8. upgrading	[ˌʌpˈgreɪdɪŋ]	v.	使升级；提升；改良品种

Notes 注释

1. automatic assembly line　　　　　　自动装配线
2. hydraulic-air pressure technology　　液压气动技术
3. single chip microcomputer technology　单片机技术
4. artificial intelligence　　　　　　　　人工智能
5. tandem joint robots　　　　　　　　串联关节机器人
6. horizontal joint robots　　　　　　　水平关节机器人
7. Delta parallel joint robots　　　　　Delta 并联关节机器人
8. work frequency　　　　　　　　　　工作频率
9. consumer electronics　　　　　　　　消费类电子产品
10. home appliances　　　　　　　　　家用电器
11. aerospace industry　　　　　　　　航空航天工业

12. accurate positioning　　　　　　　　准确定位
13. automatic logistics transport　　　　自动物流运输

Reference Translation　参考译文

　　搬运机器人是一种可以进行自动化搬运作业的工业机器人。最早的搬运机器人出现在 1960 年的美国，Versatran 和 Unimate 两种机器人首次用于搬运作业。搬运作业是指用一种设备握持工件，并将工件从一个加工位置移动到另一个加工位置。搬运机器人可以以安装不同的末端执行器以完成各种不同形状和状态的工件搬运工作，大大减轻了人类繁重的体力劳动。目前，世界上使用的搬运机器人逾 10 万台，被广泛应用于机床上下料、冲压机自动化生产线、自动装配流水线、码垛搬运、集装箱等的自动搬运。部分发达国家已制定出人工搬运的最大限度，超过限度的必须由搬运机器人来完成。

　　搬运机器人是近代自动控制领域出现的一项高新技术，涉及力学、机械学、液压气压技术、自动控制技术、传感器技术、单片机技术和计算机技术等学科领域，已成为现代机械制造生产体系的重要组成部分。它的优点是可以通过编程来完成各种预期的任务，在自身结构和性能上具有人和机器的优势，尤其体现了人工智能和适应性。

　　常见的搬运机器人有串联关节机器人、水平关节机器人、Delta 并联关节机器人和 AGV 搬运机器人等。以上各类机器人在搬运行业中各有特点。

　　其中，串联关节机器人拥有 4 个或 6 个旋转轴，类似于人类的手臂，更加灵活，负载较大，多用于较重物品的搬运。

　　SCARA 机器人是一种水平多关节机器人，工作频率更快，特别适合实验室自动化、医药、消费电子、食品、汽车、电子配件、PC 外设、半导体、塑料、家电和航空航天工业中物品和工件的搬运。

　　Delta 机器人属于高速、轻载的并联机器人。Delta 机器人主要应用于食品、药品和电子产品等的搬运与装配。Delta 机器人以其重量轻、体积小、运动速度快、定位精确、成本低、效率高等特点，在市场上得到了广泛应用。

　　我国正处于工业转型升级的关键时期，越来越多的企业在生产制造中引进和使用了工业机器人，这将深刻影响中国制造的方方面面。

Welding Robot
焊接机器人

　　The welding robot is an industrial robot engaged in welding. Industrial robot is a versatile, reproducible manipulator with three or more programmable axes for industrial automation. In order to adapt itself to different uses, the mechanical interface of the robot's last axis is usually a connecting flange, which can be attached to different tools or end effectors. The welding robot is the industrial robot of which a welding tong or torch is in the end, so that it can be weld, cut or hot spray.

Video 49

Unit 13　Industry Application of Robot
机器人行业应用

With the development of electronic technology, computer technology, numerical control and robot technology, since the beginning of the 1960s, automatic welding robot began to be used for production; its technology has become increasingly mature. It has mainly the following advantages:

1) To stabilize and improve the quality of welding, welding quality can be reflected in the form of value.

2) To improve labor productivity.

3) To improve the labor intensity of workers, can work in a harmful environment;

4) To reduce requirements for workers' operating techniques.

5) To shorten the product replacement of the preparation cycle, reduce the corresponding equipment investment.

Therefore, welding robot has been widely used.

Welding robot mainly includes two parts——robot and welding equipment. The robot consists of the robot body and the control cabinet(hardware and software). Take arc welding as an example, the welding equipment consists of the welding power(including its control system), wire feeder, welding gun and other components. For intelligent robot it should also have a sensing system, such as laser or camera sensors and their control devices.

In addition, if the workpiece in the entire welding process is without displacement, you can use the fixture to locate the workpiece on the table; this system is the simplest. However, in the actual production, many workpieces in the welding need to change position, so that the weld in a better position(posture) is welded. For this situation, the positioning machine and the robot can separately move, that is the robot welds again after the displacement machine changing position; and they can also be moving at the same time, the displacement machine is changing position while the robot is welding, that is often said that the displacement machine and robot coordinate to move. At this time the movement of the positioning machine and the movement of the robot are compound movement, so that the movement of the welding gun relative to the workpiece can meet the weld trajectory, and can meet the requirements of the welding speed and the welding gun's posture. In fact, the shaft of the positioner has become part of the robot, and this welding robot system can be up to 7-20 axes or more.

Vocabulary　词汇

1. torch	[tɔːtʃ]	n.	火把；火炬　v. 像火炬一样燃烧
2. stabilize	['steɪbɪlaɪz]	v.	使稳固，使安定　v. 稳定，安定
3. displacement	[dɪs'pleɪsmənt]	n.	取代；位移；排水量
4. component	[kəm'pəʊnənt]	n.	部件；组件；成分
5. device	[dɪ'vaɪs]	n.	装置；策略；图案
6. fixture	['fɪkstʃə(r)]	n.	设备；固定装置
7. workpiece	['wɜːkpiːs]	n.	工件；轧件；工件壁厚
8. posture	['pɒstʃə]	n.	姿势；态度　v. 摆姿势
9. separately	['sepərɪtli]	adv.	分别地；分离地；个别地
10. coordinate	[kəʊ'ɔːdɪnɪt]	adj.	并列的；同等的　vt. 调整；整合

11. shaft　　　　　　　[ʃæft]　　　　　　　n. [机]轴；箭杆　v. 利用

Notes 注释

1. end effector　　　　　　　末端执行器
2. welding tong　　　　　　　焊钳
3. hot spray　　　　　　　热喷雾
4. electronic technology　　　　　　　电子技术
5. numerical control　　　　　　　数控
6. labor productivity　　　　　　　劳动生产率
7. labor intensity　　　　　　　劳动强度
8. preparation cycle　　　　　　　准备周期
9. welding equipment　　　　　　　焊接设备
10. arc welding　　　　　　　电弧焊
11. wire feeder　　　　　　　送丝机；送丝装置
12. welding gun　　　　　　　焊枪
13. intelligent robot　　　　　　　智能机器人
14. sensing system　　　　　　　传感系统
15. camera sensor　　　　　　　相机传感器
16. positioning machine　　　　　　　定位机
17. displacement machine　　　　　　　变位机
18. compound movement　　　　　　　复合运动
19. weld trajectory　　　　　　　焊接轨迹

Reference Translation 参考译文

焊接机器人是一种从事焊接作业的工业机器人。工业机器人是一种多用途的、可重复编程的自动控制操作机（manipulator），具有三个或更多可编程的轴，用于工业自动化领域。为了适应不同的用途，机器人最后一个轴的机械接口通常是一个连接法兰，可接装不同工具（或称为末端执行器）。焊接机器人就是在工业机器人的末轴法兰装接焊钳或焊（割）枪的，使之能进行焊接、切割或热喷涂等作业。

随着电子技术、计算机技术、数控及机器人技术的发展，自动焊接机器人从20世纪60年代开始用于生产以来，其技术已日趋成熟，主要有以下优点：

1）为了稳定和提高焊接质量，可以将焊接质量以数值的形式反映出来。

2）提高了劳动生产率。

3）改善了工人的劳动强度，而且能够在有害环境下工作。

4）降低了对工人操作技术的要求。

5）缩短了产品改型换代的准备周期，减少了相应的设备投资。

因此，焊接机器人得到了广泛的应用。

焊接机器人主要包括机器人和焊接设备两部分。机器人由机器人本体和控制柜（硬件及软件）组成。而焊接设备，以弧焊为例，则由焊接电源（包括其控制系统）、送丝机、焊枪等部分组成。对于智能机器人还应有传感系统,如激光或摄像传感器及其控制装置等。

另外,如果工件在整个焊接过程中无须变位,就可以用夹具把工件定位在工作台上,这种系统是最简单的。但在实际生产中,许多工件在焊接时需要变位,使焊缝处在较好的位置(姿态)下进行焊接。对于这种情况,定位机与机器人可以是分别运动,即变位机变位后机器人再焊接;也可以是变位机变位的同时机器人进行焊接,也就是常说的变位机与机器人协调运动。这时变位机的运动及机器人的运动复合,使焊枪相对于工件的运动既能满足焊缝轨迹又能满足焊接速度及焊枪姿态的要求。实际上这时变位机的轴已成为机器人的组成部分,这种焊接机器人系统可以多达 7~20 个轴,或更多。

Part 3　Painting Robot
喷涂机器人

If your company has a painting application, why not automate it with a robot? Painting automation is a simpler, safer and better method than to any manual painting process. Furthermore, industrial painting robots are much more accessible than they used to be. Not only are there more models on the market, but also, they are more affordable than ever before.

Video 50

Then what would be like if all go with automation? It is about the cost of doing business and how to decrease that cost. Well, if you're painting, the way to decrease costs across the board is to automate!

1. The Cost of Time

The expression "time is money" definitely applies to painting and coating jobs. Manual painting costs more because it takes longer. Not only does a worker operate more slowly than a painting robot, but workers also have to be allowed breaks, lunches and vacations. The repetitive nature of the painting task can also cause fatigue, stress, and injury. Robots, on the other hand, are capable of painting 24 hours a day, 365 days a year. They work efficiently no matter how long they have been running, increasing throughput, while never decreasing quality.

2. The Cost of Materials

When a worker is painting and coating manually, mistakes like overspray can waste materials and decrease the quality of products, possibly damaging them. The quality of manual painting is never consistent. Overall, it is a messy process that will end up costing your company more money.

On the other hand, robots conserve paint and work with incredible precision and consistency. Typical paint savings when using robotic automation is 15-30 percent. Since they are programmed to spray the same amount of material on every product, they have fewer overspray problems and create a consistent coating on each part every time.

3. The Cost of Safety

Paint contains hazardous materials like Xylene and Toluene. These substances are common in several different paint and coating materials, and it is very dangerous for humans to work around them for long period of time because of the toxic fumes the chemicals produce. Instead, you can choose to automate with painting robots, which control and isolate the above hazards. Workers are removed from the dangers and placed in supervisory roles.

What is a painting robot?

"Painting Robot" is an industry term for a robot that has two major differences from all other standard industrial robots.

1) Explosion-proof arms. Painting robots are built with pxplosion-proof robot arms, meaning that they are manufactured in such a way that they can safely spray coatings that create combustible gasses. Usually these coatings are solvent based paints, when applied, it must create an environment that must be monitored for fire safety.

2) Self-contained paint systems. When painting robots were first designed, they only had one function-to work safely in a volatile environment. As acceptance and use expanded, painting robots grew into a unique subset of industrial robots, not just a traditional robot with pxplosion-proof options. Painting robots now have the ability to control all aspects of spray parameters. Fan air, atomization air, fluid flow, voltage, etc., can all be controlled by the robot control system.

The ABB painting robot is shown in Figure 13-1.

Figure 13-1　ABB Painting Robot

Vocabulary　词汇

1. manual	[ˈmænjuəl]	adj. 手工的；体力的　n. 手册，指南
2. model	[ˈmɒdl]	n. 模型；典型；模范；模特儿；样式
3. automate	[ˈɔːtəmeɪt]	v. 自动化，使自动化
4. fatigue	[fəˈtiːg]	n. 疲劳，疲乏　v. 疲劳，使疲劳　adj. 疲劳的
5. throughput	[ˈθruːˈpʊt]	n. 生产量，生产能力
6. overspray	[ˈəʊvəspreɪ]	n. 超范围喷涂　v. 过喷
7. consistency	[kənˈsɪstənsi]	n. [计] 一致性；稠度；相容性
8. spray	[spreɪ]	n. 喷雾；喷雾器　v. 喷射，喷
9. substance	[ˈsʌbstəns]	n. 物质；实质；资产；主旨
10. isolate	[ˈaɪsəleɪt]	v. 使隔离；使孤立　adj. 隔离的
11. supervisory	[ˈsjuːpəˈvaɪzəri]	adj. 监督的
12. solvent	[ˈsɒlvənt]	adj. 有偿付能力的；有溶解力的　n. 溶剂；解决方法

Unit 13　Industry Application of Robot
机器人行业应用

13. monitor	['mɒnɪtə]	v.	监督；监控，监听；测定；
14. volatile	['vɒlətaɪl]	adj.	[化学] 挥发性的；不稳定的
		n.	挥发物；有翅的动物
15. subset	['sʌbset]	n.	[数] 子集；子设备；小团体
16. parameter	[pə'ræmɪtə(r)]	n.	系数；参量
17. voltage	['vəʊltɪdʒ]	n.	[电] 电压

Notes　注释

1. painting robot —— 喷涂机器人
2. painting application —— 喷涂应用
3. painting automation —— 喷涂自动化
4. manual painting —— 手工喷涂
5. industry term —— 行业术语
6. explosion proof arms —— 防爆手臂
7. combustible gas —— 可燃气体
8. self-contained paint systems —— 独立的油漆系统
9. atomization air —— 雾化空气
10. fluid flow —— 流体流动

Reference Translation　参考译文

如果您的公司有喷涂作业，为什么不用机器人来实现自动化呢？喷涂自动化是一种比任何手工喷涂过程更简单、更安全和更优越的方法。此外，工业喷涂机器人比以前更容易获得。市场上不仅有更多的型号，而且比以往更加实惠。

喷涂实现自动化后会怎么样呢？这就涉及成本以及如何降低成本了。那么，如果你正在喷涂，降低成本的有效方式就是自动化！

1. 时间成本

"时间就是金钱"的说法绝对适用于喷涂和涂装工作。手工喷涂成本更高，耗时更长。工人不仅比喷涂机器人干得慢，而且他们还不得不休息、就餐和休假。喷涂任务的重复性也可能导致疲劳、压力和伤害。另一方面，机器人能够每天工作24h，每年工作365天。无论运行多长时间，它们都能够在提高工作量的同时不降低工作质量，并且能够高效地工作。

2. 原料成本

当工作人员手工涂装和喷涂时，过度喷涂这种错误会浪费材料，降低产品质量，甚至可能损坏产品。手工喷涂的质量是从来都不一致的。总的来说，这是一个非常麻烦的过程，最终导致您的公司花费更多的金钱。另一方面，喷涂机器人保证油漆和工作方面具有令人难以置信的精度和一致性。使用机器人自动化喷涂时，典型的油漆节省率可达15%~30%。由于它们被编程设定在每个产品上喷洒相同数量的材料，所以它们很少有过喷问题，并且每次在每个部件上形成一致的涂层。

3. 安全成本

油漆中含有有害物质，如二甲苯和甲苯。这些物质在不同的喷涂和涂层材料中是常见的，并且由于化学物质产生的有毒烟雾，人类长期在其中工作是非常危险的。这时，您可以选择自动化喷涂机器人，它可以控制和隔离上述危险。工人可以从危险中脱离出来，担任监督工作质量的角色。

什么是喷涂机器人？

"喷涂机器人"是机器人的行业术语，与其他标准工业机器人有两个主要区别：

1）防爆手臂。喷涂机器人是用防爆机器人手臂制造的，这意味着以这种方式制造的它们可以安全地喷涂那些产生可燃气体的涂料。通常这些涂料是溶剂型涂料，当被喷涂时，需要创造一个监测消防安全的环境。

2）独立涂料系统。当喷涂机器人首次设计时，它们只具有一个功能——在易挥发环境中安全工作。随着接受和使用范围的扩大，喷涂机器人逐渐成为工业机器人的一个独特分支，而不仅仅是具有防爆功能的传统机器人。喷涂机器人现在有能力控制喷涂参数的各个方面。例如：风机空气、雾化空气、流体流量、电压等都可以由机器人控制系统控制。

Polishing Robot
打磨机器人

"Polishing Robot" is an industry term for robots that can polish automatically, it is widely used in the fields of 3C, sanitary ware, IT, auto parts, industrial parts, medical devices and civil products, as shown in Figure 13-2.

Figure 13-2　Application of Polishing Robot in Auto Industry　　　　Video 51

Robotic polishing is the process of refining surface until they are smooth and shiny. This application is repetitive and tedious while requiring extreme consistency. Polishing robots are programmed to apply the appropriate pressure and move precisely in the right direction, for consistent, thorough, high-quality products.

Unit 13 Industry Application of Robot

机器人行业应用

1. The Advantages of Robotic Material Removal

With the flexibility, repeatability and extreme precision, polishing robots are possible to grind, trim or polish almost any material to achieve a consistent high-quality finish. These robots also improve production time while reducing waste. Robots save workers from both the drudgery and safety hazards associated with polishing. Polishing robots are unharmed by fumes and dust. Besides, robotic polishing is better for the environment because dry abrasive wheels are used instead of chemical solutions.

2. Methods for Handling the Process with a Robot

Currently, polishing robots are mostly the six-axis robots in industrial applications. Based on the different properties of the end effector, there are the two main approaches for the robot—to hold the workpiece or the tool, as shown in Figure 13-3.

a) Holding the Workpiece   b) Holding the Tool

Figure 13-3 Classifications of Polishing Robot

The polishing robot holding the workpiece is usually used to polish the relatively small workpiece. It grabs the workpiece without being polished by its end effector and makes this workpiece polished on the polishing machine. In addition, it is possible to add value to the system by letting the robot unload the finished part onto a conveyor or similar equipment. There is usually one or several tools around the polishing robot. However, robot holding the tool generally applies to large parts or workpieces that are heavy for the polishing robot. The handling of the workpiece can be done manually, and the robot automatically changes the required polishing tools from the tool rack. Usually applying the force control device in this system to ensure that the polishing pressure between the tool and the workpiece is consistent and to compensate the consumption of polishing head. The polishing quality can be uniformed by the force control device while the teach also can be simplified.

In practical application, it is also possible to have several robots working together for ultimate flexibility. One robot holds the part, while others manipulate the tool.

3. The Structure of Polishing Robot's System

The polishing robot's system of holding the tool consists mainly of manipulator, controller, teach pendant, the operating system of polishing and peripheral equipment, as shown in Figure 13-4.

(1) The operating system of polishing　This system includes polishing power head, inverter, force sensor, force sensor controller and automatic tool changer(ATC).

(2) Peripheral equipment　The peripheral equipment of polishing robot's system, including fence, the robot platform, transmission device, the device to put workpiece, quieter, facilitates the system to complete the whole polishing.

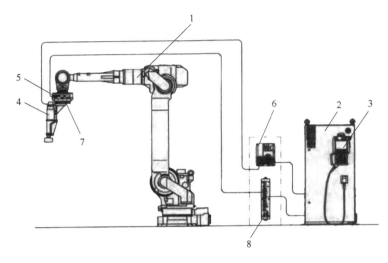

Figure 13-4　The Structure of Polishing Robot's System
1—Manipulator　2—Controller　3—Teach pendant　4—End effector　5—Force sensor　6—Inverter
7—Automatic tool changer(ATC)　8—Force sensor controller

Vocabulary　词汇

1. civil	['sɪvəl]	adj.	公民的；民间的；文职的
2. grind	[ɡraɪnd]	v.	磨碎；磨快；折磨
3. trim	[trɪm]	v.	修剪；整理；装点；削减
4. drudgery	['drʌdʒəri]	n.	苦工，苦差事
5. dust	[dʌst]	n.	灰尘；尘埃；尘土
6. hold	[həʊld]	v.	持有；保存；支持；有效
7. conveyor	[kən'veɪə(r)]	n.	输送机，[机] 传送机
8. compensate	['kɒmpenseɪt]	v.	赔偿；抵消；补偿；赔偿
9. consumption	[kən'sʌmpʃn]	n.	消费；消耗；肺痨
10. manipulate	[mə'nɪpjuleɪt]	v.	操纵；巧妙地处理；篡改
11. inverter	[ɪn'vɜːtə]	n.	换流器；[电子] 反相器

Notes　注释

1. polishing robot　　　　　　　　打磨机器人
2. sanitary ware　　　　　　　　　洁具

3. auto part   汽车零件
4. refining surface   精制面
5. extreme precision   极度精准；非常精准
6. safety hazard   安全隐患
7. abrasive wheel   砂轮
8. chemical solution   化学溶液
9. tool rack   工具架
10. the operating system of polishing   打磨操作系统
11. polishing power head   打磨头
12. force sensor   力传感器
13. force sensor controller   力传感器控制器
14. transmission device   输送装置
15. automatic tool changer   自动换刀

Reference Translation   参考译文

  打磨机器人是指可进行自动打磨的工业机器人，广泛应用于3C、卫浴、IT、汽车零部件、工业零部件、医疗器械、民用产品等行业，如图13-2所示。

  机器人抛光是精炼表面的过程，直到它们光滑发亮。这过程是重复且单调的，同时需要极端的一致性。为了得到一致、精湛、高质量的产品，打磨机器人被设定为提供适当的压力，并向正确的方向精确地移动。

1. 机器人去除材料的优点

  具有灵活性、可重复性和极高精度的打磨机器人可以研磨、修整或抛光几乎任何材料，以达到一致的高质量的成品。这些机器人还可以在减少浪费的同时缩短生产时间。机器人能让工人免于因打磨而带来的繁重工作和安全隐患。打磨机器人不会受到烟雾和灰尘的伤害。此外，机器人抛光对环境更有利，因为干燥的研磨轮代替了化学溶液。

2. 机器人打磨处理的方法

  在目前的实际应用中，打磨机器人大多数是六轴机器人。根据末端执行器的不同特性，打磨机器人夹持工件和工具的方法有两种，如图13-3所示。

  机器人夹持工件通常用于处理相对比较小的工件，机器人通过其末端执行器抓取待打磨工件并操作工件在打磨设备上进行打磨。另外，可以通过让机器人把成品卸在传送带上或类似的装置上，从而增加系统的价值。一般在该机器人的周围有一个或多个工具。但是，机器人夹持工具一般用于对大型工件或对于机器人来说比较重的工件进行加工。工件的装卸可由人工来完成，机器人自动地从工具架上更换所需的打磨工具。通常在此系统中采用力控制装置来保证打磨工具与工件之间的压力一致，补偿打磨头的消耗，获得均匀一致的打磨质量，同时也能简化示教过程。

  在实际应用中，也有可能让几个机器人协同工作以获得最大的灵活性。一个机器人控制工件，另一个机器人操纵工具。

3. 打磨机器人的系统组成

  打磨机器人持刀系统主要由操作机、控制器、示教器、打磨作业系统和周边设备组成，如图 13-4 所示。

（1）打磨作业系统　打磨作业系统包括打磨动力头、变频器、力传感器、力传感器控制器和自动快换装置等。

（2）周边设备　周边设备包括安全保护装置、机器人安装平台、输送装置、工件摆放装置、消音装置等，用以辅助打磨机器人系统完成整个装配作业。

Unit 14 Machine Vision Technology

机器视觉技术

Part 1　Introduction to Machine Vision
机器视觉概述

1. Definition

Machine vision(MV) is the technology used to provide imaging-based automatic inspection and analysis. Machine vision is used for applications such as process control and robot guidance, usually in industry.

Figure 14-1 shows an example of machine vision system used for bottle fill-level inspection at a beverage factory. Each bottle of beverage passes through an inspection sensor, triggering a vision system to take a picture of the bottle. After acquiring the image and storing it in memory, a vision software processes or analyzes the image.

If the system detects a bottle not fully filled, it signals a diverter to reject the bottle. An operator can view rejected bottles and ongoing process statistics on a display.

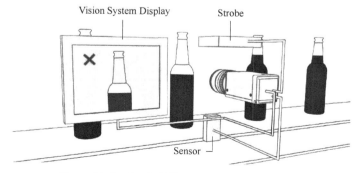

Figure 14-1　Bottle Fill-Level Inspection Example

Video 52

2. Features of Machine Vision

Human vision does well in qualitative interpretation of a complex, unstructured scene, while machine vision excels at quantitative measurement of a structured scene because of its speed, accuracy, and repeatability.

For example, on a production line, a machine vision system can inspect hundreds, or even thousands of parts per minute. A machine vision system equipped with right cameras can easily inspect object details which are too small for the human eyes to observe. Comparisons between machine vision systems and human vision systems are shown in Table 14-1.

Table 14-1 Comparisons Between Machine Vision Systems and Human Vision Systems

Characteristics	Machine Vision System	Human Vision System
Adaptability	Poor, susceptible to complex background and environmental changes	Strong, able to identify targets in complex and changeable environments
Intelligence	Poor, cannot recognize the changing target well	Strong, can recognize the changing target, and summarize the rules
Grey-level Resolution	Strong, 256 gray levels	Poor, 64 gray levels
Spatial Resolution	Strong, can observe both tiny objects and large objects	Poor, cannot observe small targets
Color Recognition	Poor, but the color can be quantified.	Strong, but can be easily influenced by people's psychology and cannot be quantified
Speed	Fast, shutter time up to 10 microseconds	Slow, unable to see faster moving targets
Accuracy	High, micron level, easy to quantify	Low, unable to quantify
Photosensitive Range	Wide, including visible light, X-ray	Narrow, visible light in the range of 400nm to 750nm
Adaptability	Strong, protective devices can also be installed	Poor, there are many environments that are harmful to people.

In removing physical contact between a test system and the parts being tested, machine vision prevents part damage and eliminates the maintenance costs associated with wear and tear on mechanical components.

Machine vision brings additional safety and operational benefits by reducing human involvement in a manufacturing process. Moreover, it prevents human contamination of clean rooms and protects human workers from hazardous environments.

Vocabulary 词汇

1. inspection   [ɪnˈspekʃn]   n. 检查
2. trigger   [ˈtrɪɡə(r)]   v. 引起；触发
3. cap   [kæp]   n. 瓶盖
4. diverter   [daɪˈvɜːtə]   n. 分流器

5. qualitative	['kwɒlɪtətɪv]	adj. 定性的
6. interpretation	[ɪn,tɜːprə'teɪʃn]	n. 理解；解释；说明
7. excel	[ɪk'sel]	v. 擅长；善于
8. quantitative	['kwɒntɪtətɪv]	adj. 定量的
9. repeatability	[rɪ,piːtə'bɪlɪti]	n. 重复性；可重复性
10. part	[pɑːt]	n. 工件
11. adaptability	[ə,dæptə'bɪlɪti]	n. 适应性；可用性；灵活性；适应能力
12. susceptible	[sə'septəbl]	adj. 易受影响（或伤害等）的
13. photosensitive	[,fəʊtəʊ'sensətɪv]	adj. 感光的；光敏的
14. involvement	[ɪn'vɒlvmənt]	n. 参与；加入
15. contamination	[kən,tæmɪ'neɪʃn]	n. 污染
16. hazardous	['hæzədəs]	adj. 危险的；有害的

Notes 注释

1. machine vision —— 机器视觉
2. wear and tear —— 磨损
3. automatic inspection —— 自动检测
4. process control —— 过程控制
5. robot guidance —— 机器人引导
6. inspection sensor —— 检测传感器
7. unstructured scene —— 非结构化场景
8. structured scene —— 结构化场景
9. grey-level resolution —— 灰度分辨率
10. spatial resolution —— 空间分辨率
11. color recognition —— 颜色识别
12. photosensitive range —— 感光范围
13. physical contact —— 物理接触

Reference Translation 参考译文

1. 定义

机器视觉（MV）是一种用于提供基于图像的自动检测和分析的技术。机器视觉可用于工业过程控制和机器人引导等应用。

图 14-1 显示了一个用于饮料厂装瓶检测的机器视觉系统示例。每瓶饮料都要经过一个检测传感器，它会触发一个视觉系统对瓶子拍照。在获取图像并将其存储在内存中之后，视觉软件对图像进行处理或分析。

如果系统检测到一个没有被装满的瓶子，它会向分流器发出信号，拒绝瓶子。操作员可以在显示屏上查看被拒绝的瓶子和正在进行的统计过程。

2. 机器视觉的特点

人类视觉适合对复杂、非结构化场景进行定性解释，而机器视觉因其速度、准确性和可重复性而擅长对结构化场景进行定量测量。

例如，在一条生产线上，机器视觉系统每分钟可以检查数百个甚至数千个零件。由装载了合适相机的机器视觉系统轻松地检查人眼无法看到的物体细节。机器视觉系统与人类视觉系统的对比见表14-1。

表14-1 机器视觉系统与人类视觉系统的对比

性能特征	机器视觉系统	人类视觉系统
适应性	差，容易受复杂背景及环境变化影响	强，可在复杂多变的环境中识别目标
智能性	差，不能很好地识别变化的目标	强，可识别变化的目标，能总结规律
灰度分辨力	强，一般为256灰度级	差，一般只能分辨64个灰度级
空间分辨力	强，可以观测很微小或很大的目标	较差，不能观看微小的目标
色彩识别能力	较差，但可以量化	强，易受人的心理影响，不能量化
速度	快，快门时间可达到10μs	慢，无法看清较快运动的目标
观测精度	高，可到微米级，易量化	低，无法量化
感光范围	较宽，包括可见光、X光等	400~750nm范围的可见光
环境适应性	强，还可以加装防护装置	差，且有许多场合对人有害

机器视觉系统可进行非接触检测，即消除了测试系统和被测试零部件之间的物理接触时，机器视觉可防止零部件损坏，并去除与机械部件磨损相关的维护成本。

机器视觉通过减少人工参与制造过程带来额外的安全和操作效益。此外，它还防止了洁净室的人为污染，并保护工人免受危险环境的影响。

Part 2 Technical Basis of Machine Vision
机器视觉技术基础

1. Image Formation

Image is the projection of a space object on the image plane, through an imaging system. The gray level of each pixel in the image reflects the intensity of the reflected light of the point on the object surface, and the position of the point on the image is related to the geometric position of the corresponding point on the object surface.

Video 53

Machine vision calculates the geometric parameters of the measured object in three-dimensional space based on the camera imaging model. Therefore, the establishment of a suitable camera imaging model is an important step in three-dimensional measurement.

Let us consider a lens with radius R and refractive index n. We assume that this lens is surrounded by air, with an index of refraction equal to 1, and that it is thin.

Consider a point P located at (negative) depth z off the optical axis and denote by PO the ray passing through this point and the center O of the lens (Figure 14-2). It follows from Snell's Law that the ray PO is not refracted and that all other rays passing through P are focused by the thin lens on the point p with depth along PO.

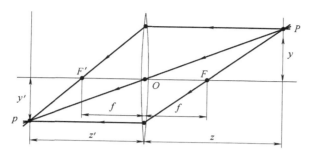

Figure 14-2 A Thin Lens

In this case, lens imaging model can be approximated by pinhole imaging model. The pinhole model assumes that the reflected light from the surface of an object passes through a pinhole and projects onto the image plane, which satisfies the straight-line propagation of light.

The pinhole model is mainly composed of optical center, imaging plane and optical axis, as shown in Figure 14-3. Pinhole model and lens imaging model have the same imaging principle, that is, the image point is the intersection of the line, which passes object point and light center, and the image plane.

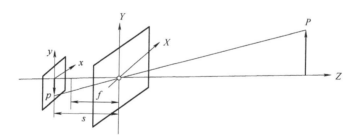

Figure 14-3 Pinhole Perspective Model

2. Distortion

In geometric optics, distortion is a deviation of actual projection from rectilinear projection. It is a form of optical aberration.

Although distortion can be irregular or follow many patterns, the most commonly encountered distortions are radially symmetric, or approximately so, arising from the symmetry of a photographic lens. These radial distortions can usually be classified as either barrel distortions or pincushion distortions.

(1) Barrel distortion   In barrel distortion, as shown in Figure 14-4a, image magnification decreases with distance from the optical axis. The apparent effect is that of an image which has been mapped around a barrel. In a zoom lens, barrel distortion appears in the middle of the lens's focal length range and is worst at the wide-angle end of the range.

(2) Pincushion distortion   In pincushion distortion, as shown in Figure 14-4b, image magnification increases with the distance from the optical axis. The visible effect is that lines that do not go

through the center of the image are bowed inwards, towards the center of the image, like a pincushion.

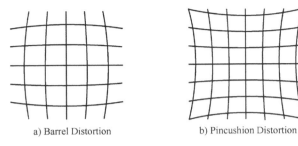

a) Barrel Distortion   b) Pincushion Distortion

Figure 14-4   Radial Distortion

3. Image Processing Technology

Image processing is a method to perform some operations on an image, in order to get an enhanced image or to extract some useful information from it. It is a type of signal processing in which input is an image and output may be image or characteristics/features associated with that image.

Nowadays, image processing is among rapidly growing technologies. In this section, we will introduce two basic image processing technology, filtering and template matching.

(1) Filtering   Filtering is a technique for modifying or enhancing an image. Filtering is a neighborhood operation, in which the value of any given pixel in the output image is determined by applying some algorithm to the values of the pixels in the neighborhood of the corresponding input pixel. A pixel's neighborhood is some set of pixels, defined by their locations relative to that pixel. In this section, we will introduce two image filtering techniques with wide application in machine vision.

1) Gaussian blur: In image processing, a Gaussian blur is the result of blurring an image by a Gaussian function. It is a widely used effect in graphics software, typically to reduce image noise and reduce detail. Figure 14-5 shows an example of applying Gaussian filter in an image.

a) Original Image   b) Gaussian Blurred Image

Figure 14-5   Gaussian Blur Effect

2) Image Sharpening: The purpose of sharpening is to enhance the contrast of gray level, since

the edges and contours are located in the places where the gray level changes.

Image sharpening and edge detection are very similar. First, the edge is found, then the edge is added to the original image, which strengthens the edge of the image and makes the image look sharper. Figure 14-6 shows an example of image sharpening.

a) Original Image b) Sharpened Image

Figure 14-6 Image Sharpening Effect

(2) Template matching Template matching is a technique in digital image processing for finding small parts of an image which match a template image. It can be used in manufacturing as a part of quality control, a way to navigate a mobile robot, or as a way to detect edges in images.

A simple template matching method is to move the template image in the image to be searched and measure the gray level difference between each pixel of the sub-image and the template image. If the sum of differences is below certain threshold, the template is found and the coordinate of the sub-image is returned as the matching position.

Take Figure 14-7 as an example to illustrate the processing of template matching. Figure 14-7a is the image to be searched, and Figure 14-7b is the template image. Template matching is to find the pentagonal star in Figure 14-7b in Figure 14-7a. The matching result is shown in Figure 14-7c.

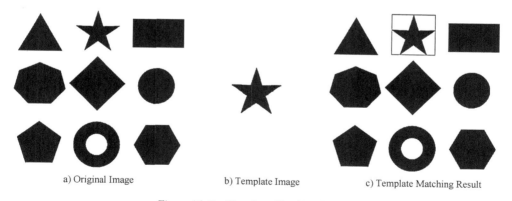

a) Original Image b) Template Image c) Template Matching Result

Figure 14-7 Template Matching Result

Digital image processing is the basis for further image recognition, analysis and understanding. In this part, we introduced three basic methods of digital image processing, Gaussian blur, image sharpening and template matching. Beside these methods, there are a large number of methods, which can realize various image processing functions. For example, the image segmentation algorithm can separate the foreground from the background of the image, and the edge detection algorithm can extract the edge information of the image and so on.

Vocabulary 词汇

1. projection [prəˈdʒekʃn] n. 投影
2. geometric [ˌdʒiːəˈmetrɪk] adj. 几何（学）的
3. establishment [ɪˈstæblɪʃmənt] n. 建立，创建；企业，科研机构
4. lens [lenz] n. 透镜
5. ray [reɪ] n. 光线
6. refract [rɪˈfrækt] v. 使（光线）折射
7. approximate [əˈprɒksɪmət] v. 近似；接近
8. intersection [ˌɪntəˈsekʃn] n. 交点
9. invert [ɪnˈvɜːt] v. （使）倒转，颠倒，倒置
10. perspective [pəˈspektɪv] n. 透视法
11. distortion [dɪˈstɔːʃn] n. 失真；变形
12. deviation [ˌdiːviˈeɪʃn] n. 偏离；违背；偏差
13. rectilinear [ˌrektɪˈlɪniə(r)] adj. 直线的
14. aberration [ˌæbəˈreɪʃn] n. 像差
15. filter [ˈfɪltə(r)] n. 滤波器；v. 滤波
16. contour [ˈkɒntʊə(r)] n. 外形；轮廓
17. strengthen [ˈstreŋθn] v. 加强；增强
18. navigate [ˈnævɪɡeɪt] v. 导航

Notes 注释

1. image formation   成像
2. imaging model   成像模型
3. refractive index   折射率
4. focal length   焦距
5. pinhole imaging model   小孔成像模型
6. straight-line propagation   直线传播
7. optical center   光心
8. imaging plane   成像平面
9. optical axis   光轴
10. radial distortion   径向畸变
11. barrel distortion   桶形畸变
12. pincushion distortion   枕头畸变
13. Gaussian blur   高斯模糊
14. image sharpening   图像锐化
15. template matching   模板匹配

Unit 14　Machine Vision Technology
机器视觉技术

Reference Translation　参考译文

1. 成像原理

图像是空间物体通过成像系统在像平面上的投影。图像上每一个像素点的灰度反映了空间物体表面上点的反射光强度，而该点在图像上的位置则与空间物体表面上对应点的几何位置有关。

机器视觉是根据摄像机成像模型利用所拍摄的图像来计算三维空间中被测物体的几何参数的，因此建立合适的摄像机成像模型是三维测量中的重要步骤。

假设有一个半径为 R 和折射率为 n 的凸透镜。我们假设该透镜被空气包围，折射率等于 1，并且很薄。

假设有一个位于光轴深度为 $-z$ 处的点 P，并用 PO 表示穿过该点和透镜的中心 O 的光线（见图 14-2）。根据斯涅尔定律，光线 PO 不发生折射，通过 P 的所有光线都由薄透镜聚焦在点 p 上。

这时可以将透镜成像模型近似地用针孔成像模型来代替。针孔模型假设物体表面的反射光都经过一个针孔而投影到像平面上，即满足光的直线传播条件。

针孔模型主要由光学中心、成像平面和光轴组成，如图 14-3 所示。针孔模型与透镜成像模型具有相同的成像关系，即像点是物点和光心的连线与图像平面的交点。

2. 畸变

在几何光学中，畸变是实际投影与直线投影的偏差。它是一种光学像差。

尽管畸变可能是不规则的，也可能遵循许多模式，但最常见的畸变是径向对称的，或近似于径向对称的，这是由于摄影镜头的对称性造成的。这些径向畸变通常可分为桶形畸变或枕形畸变。

（1）桶形畸变　在桶形畸变中（见图 14-4a），图像放大率随着与光轴距离的增加而减小。桶形畸变表现为围绕一个桶映射的图像。在变焦镜头中，筒形失真出现在镜头焦距范围的中间，在广角范围的末端最严重。

（2）枕形畸变　在枕形畸变中（见图 14-4b），图像放大率随与光轴距离的增加而增大。枕形畸变可见的效果是不穿过图像中心的线条向内弯曲，朝向图像的中心，就像一个枕头。

3. 图像处理技术

图像处理是对图像执行某些操作的一种方法，以获得增强图像或从中提取一些有用的信息。它是一种信号处理操作，输入的是图像，其输出可以是图像或与该图像相关的特征或特点。

目前，图像处理技术正在迅速发展。下面介绍两种常用的图像处理技术，即图像滤波和模板匹配。

（1）滤波　滤波是一种修改或增强图像的技术。滤波是一种邻域运算，通过对相应输入像素邻域中的像素值应用某种算法来确定输出图像中任何给定像素的值。一个像素的邻域是一组像素，由它们相对于该像素的位置来定义。这里，我们将介绍两种在机器视觉中有广泛应用的图像滤波技术。

1）高斯模糊。在图像处理中，高斯模糊是用高斯函数模糊图像的结果。它是图形软件中广泛使用的一种效果，通常用于降低图像噪声和减少细节。图 14-5 显示了在图像中应用高斯滤波器的一个例子。

2) 图像锐化。图像锐化的目的是使灰度反差增强,因为边缘和轮廓都位于灰度突变的地方。图像锐化和边缘检测很像,首先找到边缘,然后把边缘添加到原来图像的上面,这样就强化了图像的边缘,使图像看起来更加清晰了。图 14-6 显示了图像锐化的一个例子。

(2) 模板匹配 模板匹配是数字图像处理中的一种技术,用于查找与模板图像匹配的图像的小部分。它可以用于制造业,作为质量控制的一部分,它也可以作为移动机器人导航的一种方法,还可以作为检测图像边缘的一种方法。

简单的模板匹配方法是:在待搜寻的图像中,移动模板图像,在每一个位置测量待搜寻图像的子图像和模板图像的像素灰度值,并计算子图像和模板图像每个像素点的灰度差值,当所有差值的和小于某个阈值时,我们认为模板被找到了,并记录其相应的位置。

以图 14-7 为例说明模板匹配的处理过程。图 14-7a 是要搜索的图像,图 14-7b 是模板图像。模板匹配是在图 14-7a 中找到五角星。匹配结果如图 14-7c 所示。

数字图像处理是进一步进行图像识别、分析和理解的基础。本节介绍了数字图像处理的三种基本方法:高斯模糊、图像锐化和模板匹配。除了这些方法外,还有很多图像处理方法,可以实现丰富的图像处理功能。例如,图像分割算法可以将前景与图像背景分离,边缘检测算法可以提取图像的边缘信息等。

Part 3 Vision System 工业机器人视觉系统

1. Composition

A typical industrial robot vision system includes components of image acquisition, processing and motion control. The vision system of industrial robot based on PC is composed of several parts as shown in Figure 14-8.

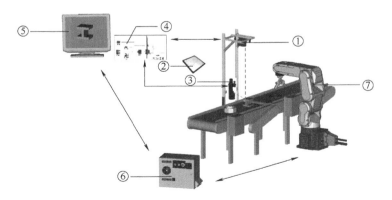

Figure 14-8 Components of a Typical Industrial Robot Vision System   Video 54

① Industrial camera and lens: this part belongs to imaging devices.
② Light source: as an auxiliary imaging device, often plays a vital role in the quality of ima-

Unit 14 Machine Vision Technology
机器视觉技术

ging.

③ Sensors: usually in the form of optical fiber switches and proximity switches, are used to judge the position and state of the object under test and to inform the image sensor to collect image correctly.

④ Image acquisition card: the image acquisition card transmits the image output from the camera to the host computer.

⑤ Image processing software: machine vision software is used to process the input image data, and then the results are obtained through certain operations.

⑥ Robot control unit: simple control is based on the use of the I/O of some image acquisition cards, while relatively complex logic/motion control must rely on additional programmable logic control unit/motion control card to achieve the necessary action.

⑦ Industrial robots and external equipment: working robots accomplish tasks such as positioning, detection, recognition, and measurement of the workpiece, according to the instructions and processing results of the control unit.

2. Working Process

Machine vision system is to transform the detected object into image signal through machine vision device and transmit it to a special image processing system. Image processing system carries out various operations on these digital signals to extract features of the target, such as area, quantity, position, length, color, and then controls the operation of field equipment.

The workflow of the machine vision system is shown in Figure 14-9. First, the light source is projected to the object under test, and the relevant image information of the object under test is captured by the CCD camera. Then, the digital signal is converted to the image processing unit through A/D(Analog to Digital) transformation. The image processing unit extracts the target features by analyzing and calculating the captured pixels, outputs the results, and finally transmits the processed related information to a PLC or robot controller.

Figure 14-9 Workflow of Machine Vision System

Vocabulary 词汇

1. auxiliary　　　[ɔːɡˈzɪliəri]　　　adj. 辅助的；备用的
2. execution　　　[ˌeksɪˈkjuːʃn]　　　n. 执行
3. workpiece　　　[ˈwɜːkpiːs]　　　n. 工件
4. position　　　[pəˈzɪʃn]　　　v. 安装；安置；使处于
5. recognition　　　[ˌrekəɡˈnɪʃn]　　　n. 识别
6. workflow　　　[ˈwɜːkfləʊ]　　　n. 工作流程

Notes 注释

1. optical fiber   光纤
2. proximity switch   接近开关
3. field equipment   现场设备
4. image capture   图像捕捉
5. image acquisition   图像采集
6. industrial camera   工业相机
7. light source   光源
8. image acquisition card   图像采集卡
9. image processing software   图像处理软件
10. robot control unit   机器人控制单元
11. external equipment   外部设备
12. field equipment   现场设备
13. feature extraction   特征提取

Reference Translation 参考译文

1. 组成

一个典型的工业机器人视觉系统包括图像获取、图像处理和运动控制等部分。基于 PC 的工业机器人视觉系统具体由图 14-8 所示的几部分组成。

① 工业相机与镜头：这部分属于成像器件。

② 光源：作为辅助成像器件，对成像质量的好坏往往能起到至关重要的作用。

③ 传感器：通常以光纤开关、接近开关等形式出现，用以判断被测对象的位置和状态，并通知图像传感器正确采集图像。

④ 图像采集卡：图像采集卡把相机输出的图像输送给计算机主机。

⑤ 图像处理软件：机器视觉软件用来完成对输入图像数据的处理，然后通过一定的运算得出结果。

⑥ 机器人控制单元：简单的控制可以直接利用部分图像采集卡自带的 I/O，相对复杂的逻辑/运动控制则必须依靠附加可编程逻辑控制单元/运动控制卡来实现必要的动作。

⑦ 工业机器人和外部设备：机器人根据控制单元的指令和处理结果，完成工件的定位、检测、识别和测量等任务。

2. 工作过程

机器视觉系统是指通过机器视觉装置将被检测目标转换成图像信号，并传送给专用的图像处理系统。图像处理系统对这些数字信号进行各种运算来抽取目标的特征，如面积、数量、位置、长度和颜色等，进而控制现场设备的作业。

机器视觉系统的工作流程如图 14-9 所示。首先将光源投射到被测物体上，通过 CCD 相机捕捉并获取被测目标的相关图像信息，然后通过 A/D 转换成数字信号传给图像处理单元，图像处理单元对捕捉到的像素进行分析运算来提取目标特征，输出结果，最后把处理过的相关信息传输给 PLC 或机器人控制器。

Unit 14 Machine Vision Technology
机器视觉技术

Part 4 Industry Application of Machine Vision
机器视觉工业应用

Machine vision application in industrial field is mainly through industrial robots. The visual functions of industrial robots can be divided into four categories, as shown in Figure 14-10.

➢ Guidance- Position and orientation location.

➢ Inspection- Identifying defects, irregularities and other manufacturing flaws.

➢ Gauging- Measuring distances and locations to assess specification adherence.

➢ Identification- Reading codes and alphanumeric characters.

Video 55

a) Guidance

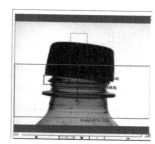
b) Inspection

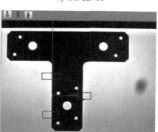
c) Gauging

d) Identification

Figure 14-10 Industry Application of Machine Vision

1. Guidance

First, machine vision systems can locate the position and orientation of a part, compare it to a specified tolerance, and ensure it is at the correct angle to verify proper assembly. Next, guidance can be used to report the location and orientation of a part in 2D or 3D space to a robot or machine controller, allowing the robot to locate the part or the machine to align the part.

2. Inspection

A machine vision system for inspection detects defects, contaminants, functional flaws, and other irregularities in manufactured products. Examples include inspecting tablets of medicine for flaws, displays to verify icons or confirm pixel presence, or touch screens to measure the level of backlight contrast.

3. Gauging

A machine vision system for gauging calculates the distances between two or more points or geometrical locations on an object and determines whether these measurements meet specifications. If not, the vision system sends a fail signal to the machine controller, triggering a reject mechanism that ejects the object from the line.

4. Identification

A machine vision system for part identification and recognition reads barcodes(1-D), data matrix codes(2-D), direct part marks(DPM), and characters printed on parts, labels, and packages. An optical character recognition(OCR) system reads alphanumeric characters without prior knowledge, whereas an optical character verification (OCV) system confirms the presence of a character string. Additionally, machine vision systems can identify parts by locating a unique pattern or identify items based on color, shape, or size.

Vocabulary 词汇

1. specification	[ˌspesɪfɪ'keɪʃn]	n.	规范
2. adherence	[əd'hɪərəns]	n.	遵守
3. defect	['diːfekt]	n.	缺点；缺陷；毛病
4. gauge	[geɪdʒ]	v.	（用仪器）测量；估计；估算
5. alphanumeric	[ˌælfənjuː'merɪk]	adj.	含有字母和数字的
6. contaminant	[kən'tæmɪnənt]	n.	致污物；污染物
7. functional	['fʌŋkʃənl]	adj.	实用的；功能的；机能的
8. irregularity	[ɪˌregjə'lærəti]	n.	不规则的事物；不整齐的事物；不平整的事物
9. presence	['prezns]	n.	存在
10. barcode	['bɑːkəʊd]	n.	（商品的）条形码
11. guidance	['gaɪdns]	n.	指导；引导

Notes 注释

1. touch screen   触摸屏
2. data matrix code   二维码
3. direct part marks （DPM）   引导直接零件标记
4. optical character recognition （OCR）   光学字符识别
5. optical character verification （OCV）   光学字符验证
6. manufacturing flaw   制造缺陷
7. functional flaw   功能缺陷

Reference Translation 参考译文

机器视觉在工业领域中应用的主要通过工业机器人来实现。工业机器人的视觉功能可以分成四类，如图 14-10 所示。
- 引导——定位位置和方向。
- 检查——识别缺陷、不规则和其他制造缺陷。
- 测量——测量距离和位置，以评估规范遵守情况。
- 识别——视觉系统能够读取代码和字母数字字符。

1. 引导

首先，机器视觉系统可以定位零件的位置和方向，将其与规定的公差进行比较，并确保其角度正确，以验证装配是否正确。接下来，可以使用引导结果向机器人或机器控制器报告零件在二维或三维空间中的位置和方向，允许机器人定位零件或机器对齐零件。

2. 检查

机器视觉系统用于检测制造产品中的缺陷、污染物、功能缺陷和其他不规则情况。例如检查药片是否有缺陷，显示器是否能验证图标或确认像素的存在，触摸屏是否能测量背光对比度。

3. 测量

用于测量的机器视觉系统计算物体上两个或多个点之间的距离或几何位置，并确定这些测量是否符合规范。如果不符合规范，视觉系统会向机器控制器发送一个失败信号，触发一个拒绝机制，将对象从生产线中分离出来。

4. 识别

用于零件识别和区分的机器视觉系统可读取条形码（1-D）、数据矩阵代码（2-D）、直接零件标记（DPM），以及打印在零件、标签和包装上的字符。光学字符识别（OCR）系统在事先不知情的情况下读取字母数字字符，而光学字符验证（OCV）系统则确认字符串的存在。此外，机器视觉系统可以通过定位独特的图案或根据颜色、形状或尺寸识别物品来识别零件。

Unit 15 Intelligent Robot

智能机器人

Since the 21st century, with the rapid development of information technology, breakthroughs in cloud computing, big data and deep learning algorithms have led to the continuous progress of intelligent technologies.

Video 56

Part 1 Definition and Classification of Intelligent Robot
智能机器人的定义与分类

1. Definition

In 1956, Marvin Minsky defined intelligent machines as "Intelligent machines are able to create abstract models of their surroundings, find solutions from abstract models once they encountered problems". This definition has an important impact on the research direction of intelligent robots in the next 30 years.

In the process of developing robots which can work in unknown or uncertain environments, people gradually realize that the essence of robotics is the combination of perception, decision-making, action and interaction technology. Therefore, systems with perception, thinking, decision-making and action are called intelligent robots.

2. Characteristics

The characteristics of intelligent robots include the following aspects:

1) Autonomy-The ability to perform a specific task completely independently in a specific environment without any external control and human intervention.

2) Adaptability-Real-time identification and measurement of surrounding objects, and adjust their own parameters according to the changes in the environment.

3) Interaction - Robots can communicate with people, external environment and other robots.

4) Learnability- Robots can form and evolve new rules of motion based on self-perception of changes in the environment. Robots can act independently and deal with problems independently.

5) Collaboration- On the basis of real-time interaction, robots can achieve machine-machine collaboration and human-machine collaboration according to tasks and requirements.

3. Classification

(1) Classification by Intelligence Level

1) Interactive robot- The interactive robot communicates with the operator through the computer system. Although it has some processing and decision-making functions, it also needs external control, as shown in Figure 15-1.

a) Wrist Robot

b) Pet Dog Robot

Figure 15-1 Interactive Robot

2) Autonomous robot- Once manufactured, autonomous robots can automatically complete various anthropomorphic tasks in various environments without external control. The autonomous robot has the modules of perception, processing, decision-making and execution, and can act independently and deal with problems like human beings. Examples of autonomous robots are shown in Figure 15-2.

a) Asimo Robot

b) Simon Robot

Figure 15-2 Autonomous Robot

(2) Classification by Application

1) Intelligent industrial robot- Industrial intelligent robots can be divided into welding robots, assembly robots, painting robots, palletizing robots, handling robots and other types according to their

specific applications. Welding robots are shown in Figure 15-3.

2) Intelligent agricultural robot- An agricultural robot is a robot deployed for agricultural purposes. The main area of application of robots in agriculture today is at the harvesting stage. Emerging applications of intelligent robots in agriculture include weed control, planting seeds, harvesting, environmental monitoring and soil analysis. An example of agricultural intelligent robot is shown in Figure 15-4.

Figure 15-3   Industrial Intelligent Robot   Figure 15-4   Intelligent Robot for Agricultural Seeding

3) Intelligent educational robot- Educational robotics teaches the design, analysis, application and operation of robots, as shown in Figure 15-5. Robots include articulated robots, mobile robots or autonomous vehicles. Educational robotics can be taught from elementary school to graduate programs. Robotics may also be used to motivate and facilitate the instruction other, often foundational, topics such as computer programming, artificial intelligence or engineering design.

4) Intelligent service robot- Robot technology has not only been used in industrial and agricultural production, scientific exploration, but also been used in people's daily life. Service robot is a general term of this kind of robot. At present, intelligent service robots are mainly used in cleaning, nursing, duty, rescue, entertainment, and other occasions, such as replacing people to maintain equipment, as shown in Figure 15-6.

Figure 15-5   Intelligent Educational Robot   Figure 15-6   Service Intelligent Robot

(3) Classification by Shape

1) Bionic intelligent robot- Bionic intelligent robot refers to intelligent robot imitating a variety of organisms, daily-used items, buildings or vehicles. Some examples of bionic intelligent robots include

bionic flying fox and bionic ants, as shown in Figure 15-7.

a) Bionic flying for b) Bionic ants

Figure 15-7　Bionic Intelligent Robot

2) Humanoid intelligent robot- Robots designed and manufactured to imitate human form and behavior are humanoid robots, which generally have human-like limbs and heads, as shown in Figure 15-8.

a) Intelligent Robot-Sophia　　　　b) Intelligent Robot-Atlas

Figure 15-8　Humanoid Intelligent Robot

Vocabulary　词汇

1. abstract	['æbstrækt]	adj.	抽象的（与个别情况相对）
2. encounter	[ɪn'kaʊntə(r)]	v.	遭遇，遇到
3. essence	['esns]	n.	本质；实质；精髓
4. interaction	[ˌɪntər'ækʃn]	n.	相互影响；干扰（涉）
5. autonomy	[ɔː'tɒnəmi]	n.	自治；自治权；自主；自主权
6. collaboration	[kəˌlæbə'reɪʃn]	n.	合作；协作
7. anthropomorphic	[ˌænθrəpə'mɔːfɪk]	adj.	人格化的；拟人化的
8. agricultural	[ˌægrɪ'kʌltʃərəl]	adj.	农业的；农用的
9. bionic	[baɪ'ɒnɪk]	adj.	仿生的；能力超人的
10. humanoid	['hjuːmənɔɪd]	adj. 有人的特点的　n. 仿真机器人

11. limb　　　　　　　[lɪm]　　　　　　　　n. 肢；臂；腿

Notes　注释

1. decision-making　　　　决策
2. intelligent robot　　　　智能机器人
3. autonomous robot　　　自主型机器人
4. welding robot　　　　　焊接机器人
5. assembly robot　　　　装配机器人
6. painting robot　　　　　喷涂机器人
7. palletizing robot　　　　码垛机器人
8. handling robot　　　　　搬运机器人
9. agricultural robot　　　农业机器人
10. service robot　　　　　服务机器人

Reference Translation　参考译文

进入 21 世纪以来，随着信息技术的飞速发展，云计算、大数据、深度学习算法上的突破带动了图像识别、语音识别、自然语言处理等智能技术不断进步。

第 1 部分　智能机器人的定义与分类

1. 定义

1956 年，马文·明斯基对智能机器进行定义："智能机器能够创建周围环境的抽象模型，一旦遇到问题，便能够从抽象模型中寻找解决方法。"该定义对此后 30 年智能机器人的研究方向产生了重要影响。

在研究和开发作业于未知及不确定环境下的机器人的过程中，人们逐步认识到机器人技术的本质是感知、决策、行动和交互技术的结合，因此将具有感知、思考、决策和动作的技术系统统称为智能机器人。

2. 特点

智能机器人的特点具体体现在以下几方面：

（1）自主性　可在特定的环境中，不依赖任何外部控制，无须人为干预，完全自主地执行特定的任务。

（2）适应性　实时识别和测量周围的物体，并根据环境的变化调节自身的参数。

（3）交互性　机器人可以与人、外部环境及与其他机器人进行信息交流。

（4）学习性　机器人在自主感知环境变化的基础上，可形成和进化出新的运动规则，自主独立地行动和处理问题。

（5）协同性　在实时交互的基础上，机器人可根据任务和需求实现机机协作和人机协同。

3. 分类

（1）按智能程度分类

1）交互型机器人。机器人通过计算机系统与操作员进行人机对话，实现对机器人的控制与操作。虽然具有部分处理和决策功能，但是还需要外部控制，如图 15-1 所示。

Unit 15　Intelligent Robot
智能机器人

2）自主型机器人。在设计制作之后，机器人无须外部控制，能够在各种环境下自动完成各项拟人化任务。自主型机器人的本体上具有感知、处理、决策和执行等模块，可以像人一样独立活动和处理问题。自主型机器人示例如图 15-2 所示。

（2）按用途分类

1）智能工业机器人。智能工业机器人依据具体应用不同，通常可分为焊接机器人、装配机器人、喷涂机器人、码垛机器人和搬运机器人等多种类型。焊接机器人示例如图 15-3 所示。

2）智能农业机器人。农业机器人是一种用于农业用途的机器人。目前，机器人在农业中的主要应用领域是作物收获。智能机器人在农业中的新应用包括杂草控制、播种、收获、环境监测和土壤分析。农业智能机器人示例如图 15-4 所示。

3）智能教育机器人。教育机器人可用于教授机器人的设计、分析、应用和操作，如图 15-5 所示。机器人包括关节机器人、移动机器人或自主车辆。从小学到研究生课程都可以教授教育机器人学。机器人技术也可以用来激励和促进其他的基础性主题，如计算机编程、人工智能或工程设计等主题的教学。

4）智能服务机器人。机器人技术不仅在工农业生产、科学探索中得到了广泛应用，也逐渐渗透到人们的日常生活领域，服务机器人就是这类机器人的一个总称。目前，智能服务机器人主要应用在清洁、护理、执勤、救援、娱乐和代替人类对设备进行维护保养等场合，如图 15-6 所示。

（3）按形态分类

1）智能仿生机器人。智能仿生机器人是指模仿各种生物、生活用品、建筑物或车辆的智能机器人。仿生机器人的一些例子包括仿生飞狐和仿生蚂蚁，如图 15-7 所示。

2）智能仿人机器人。模仿人的形态和行为而设计制造的机器人就是仿人机器人，一般仿人的四肢和头部，如图 15-8 所示。

Part 2　Basic Elements of Intelligent Robots
智能机器人的基本要素

Most experts believe that intelligent robots need three elements, perception, decision-making, and action, as shown in Figure 15-9. On the basis of these three elements, intelligent robots can make decisions through perceptual assistance, put decisions into action, complete tasks independently in complex environments, and form various intelligent behaviors.

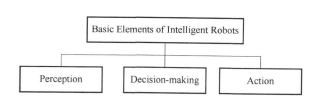

Figure 15-9　Basic Elements of Intelligent Robots

Video 57

1. Perception

Perception element includes non-contact sensors that can perceive vision, proximity and distance, and contact sensors that can perceive ability, pressure and touch.

Perception can be realized by camera, image sensor, ultrasonic transmitter, laser, conductive rubber, piezoelectric element, pneumatic element, stroke switch and other mechanical and electrical components. A robot with perception element is shown in Figure 15-10, and this robot can work together with people without the use of safety fences.

2. Decision-Making

Decision-making is the key element of the three elements. Decision-making includes judgment, logical analysis, understanding and other intellectual activities.

These intellectual activities are essentially a process of information processing, which are mainly accomplished by computers.

3. Action

For action, intelligent robots need a trackless motion mechanism to adapt to different geographic environments such as flat ground, steps, walls, stairs, ramps and so on.

Action can be accomplished by means of wheels, crawlers, legs, suckers, air cushions and other motion mechanisms. In the motion process, the real-time control of motion mechanism should be carried out, which includes not only position control, but also force control, position and force control, and expansion control. A robot with action element is show in Figure 15-11.

Figure 15-10 A Robot with Perception Element   Figure 15-11 A Robot with Action Element

Vocabulary 词汇

1. perceive	[pəˈsiːv]	v. 注意到；意识到；察觉到
2. proximity	[prɒkˈsɪməti]	n. 接近，邻近，靠近
3. laser	[ˈleɪzə(r)]	n. 激光器
4. intellectual	[ˌɪntəˈlektʃuəl]	adj. 智力的；脑力的
5. trackless	[ˈtræklɪs]	adj. 无轨的
6. geographic	[dʒɪəˈgræfɪk]	adj. 地理的
7. ramp	[ræmp]	n. 斜坡；坡道

Unit 15 Intelligent Robot
智能机器人

8. crawler	[ˈkrɔːlə(r)]	n.	爬行物（如车辆、昆虫等）
9. sucker	[ˈsʌkə(r)]	n.	吸盘
10. expansion	[ɪkˈspænʃn]	n.	扩张；膨胀

Notes 注释

1. ultrasonic transmitter　　　超声波发射器
2. conductive rubber　　　导电橡胶
3. piezoelectric element　　　压电元件
4. pneumatic element　　　气动元件
5. stroke switch　　　行程开关
6. electrical component　　　电器元件
7. air cushion　　　气垫
8. non-contact sensor　　　非接触式传感器
9. contact sensor　　　接触式传感器
10. perception element　　　感知元素
11. decision-making　　　决策
12. motion mechanism　　　运动机构
13. real-time control　　　实时控制

Reference Translation 参考译文

多数专家认为智能机器人需具备以下三个要素：感知、决策和行动，如图 15-9 所示。在这三大要素基础上，智能机器人通过感知辅助产生决策，并将决策付诸行动，在复杂的环境下自主完成任务，形成各种智能行为。

1. 感知

感知要素包括能感知视觉、接近、距离等非接触型传感器和能够感知力、压力和触觉等接触型传感器。

感知可通过摄像机、图像传感器、超声波传感器、激光器、导电橡胶、压电元件、气动元件、行程开关等来实现。一个具备感知要素的机器人如图 15-10 所示，这个机器人可以与人类共同工作，不需要安全护栏隔离。

2. 决策

决策要素是三个要素中的关键要素。决策要素包括判断、逻辑分析、理解等方面的智力活动。

这些智力活动本质上是一个信息处理过程，而计算机则是完成这个处理过程的主要手段。

3. 行动

对行动要素而言，智能机器人需要有一个无轨道运动机构，以适应诸如平地、台阶、墙壁、楼梯和坡道等不同地理环境。

它们的功能可借助轮子、履带、支脚、吸盘、气垫等运动机构来完成。在运动过程中要对运动机构进行实时控制，这种控制不仅要包括位置控制，还包括力控制、位置与力混合控制、展开控制等。一个具备行动要素的机器人如图 15-10 所示。

Application Analysis of Intelligent Robots
智能机器人应用分析

Intelligent service robots have multiple sensing systems, remote monitoring, alarm notification and multimedia navigation functions. According to the classification standards set by the International Federation of Robotics(IFR), intelligent service robots can be divided into 14 categories according to the application, as shown in Table 15-1.

Video 58

Table 15-1　Classification of Intelligent Service Robots(From IFR)

Categories	Functions
1. Domestic tasks	Vacuuming, lawn-mowing, window cleaning, pool cleaning
2. Educational entertainment	Toy/hobby robots, teaching assistance
3. Disability-assisted	Mechanical wheelchair, help turn over and bathe, etc.
4. Security	Surveillance/security robots
5. Industry	Planting and mining robots
6. Inspection purposes	Tank, tubes, pipes and sewers
7. Professional Cleaning	Floor cleaning, window and wall cleaning
8. Construction and demolition	Building construction, robots for heavy/civil construction
9. Rear Service	Mail delivery
10. Medical robotics	Robot assisted surgery or therapy, rehabilitation systems
11. Rescue & security applications	Fire and disaster fighting robots
12. Underwater systems	Maintenance of underwater pipe or detection
13. Laboratory use	Material transfer, clean laboratory
14. Public reception and guide	Exhibition guidance, restaurant reception

Medical robots mainly include minimally invasive surgical robots, rehabilitation medical robots and hospital service robots. An example surgical robot, Da Vinci surgical robot is shown in Figure 15-12.

Domestic service robots are the autonomous robot used for the household chores. Sweeping robot is used in the scenario of floor cleaning, as shown in Figure 15-13.

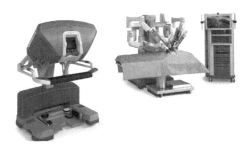

Figure 15-12　Da Vinci Intelligent Minimally Invasive Surgery System

Figure 15-13　Sweeping Robot

Unit 15 Intelligent Robot
智能机器人

Entertainment robots bring fun to our lives. They can have language ability, even singing ability, and certain ability of perception. Some example of entertainment robots are shown in Figure 15-14.

a) Pet Dog-Aibo

b) Pet Dragon-Pleo

Figure 15-14 Entertainment Robots

With the increasing application of intelligent robots, people expect that intelligent robots can serve human beings in more fields, and complete more complex tasks.

Vocabulary 词汇

1. vacuuming ['vækjuəmɪŋ] n. （用吸尘器所做的）清扫
2. surveillance [sɜːˈveɪləns] n. 监视
3. tank [tæŋk] n. （贮放液体或气体的）箱，槽，罐
4. tube [tjuːb] n. 管，管子
5. sewer [ˈsuːə(r)] n. 下水道
6. construction [kənˈstrʌkʃn] n. 建筑；建造；施工
7. demolition [ˌdeməˈlɪʃn] n. （建筑物的）摧毁，拆毁，拆除
8. therapy [ˈθerəpi] n. 治疗
9. rehabilitation [ˌriːəˌbɪlɪˈteɪʃn] n. 康复
10. reception [rɪˈsepʃn] n. 接待；招待
11. trauma [ˈtrɔːmə] n. 外伤
12. wiping [ˈwaɪpɪŋ] v. 擦；拭；抹

Notes 注释

1. International Federation of Robotics　　国际机器人联盟
2. lawn-mowing　　割草
3. remote monitoring　　远程监控
4. mining robot　　采矿机器人
5. rear service　　后勤服务
6. medical robot　　医疗机器人
7. minimally invasive surgical robot　　微创手术机器人
8. domestic service robot　　家用服务机器人

9. sweeping robot　　　　　　　　　　清扫机器人

Reference Translation　参考译文

智能服务机器人具备多种传感系统、远程监控、警报通知及多媒体导览等功能。根据国际机器人联合会（IFR）制定的分类标准，智能服务机器人可依照使用目的区分为14大类，见表15-1。

表15-1　服务用智能机器人用途分类（来源：IFR）

类　　别	功 能 项 目
1. 居家服务型	地板吸尘、除草、清洗窗户与游泳池
2. 教育娱乐型	玩具机器人、教学辅助
3. 残障辅助型	机械轮椅、协助翻身与沐浴等
4. 保安型	监控与安全机器人
5. 产业型	种植作物与开矿等用途
6. 检查用途	水箱、下水道、输油管线路检查
7. 专业清洁	清理地面、高楼外窗、外墙等
8. 建设/拆除	工地建筑、基础建设
9. 后勤处理	信件传递
10. 医疗型	辅助手术、康复使用
11. 防御救援与安全应用	火灾或灾难处理
12. 水底作业	维修水底管线或侦测
13. 实验室用	物料传递与清洁实验室环境
14. 公共接待与导览	展会导览或饭店餐厅接待

医疗机器人主要有微创外科手术机器人、康复医疗机器人和医院服务机器人，一个手术机器人的例子——达芬奇手术机器人如图15-12所示。

家政服务机器人是一种用于家庭琐事的自主机器人。扫地机器人成功应用于地面清洁，如图15-13所示。

娱乐机器人为我们的生活增添乐趣色彩，它可以具备语言能力，甚至是唱歌能力，并且有一定的感知能力。娱乐机器人的一些例子如图15-14所示。

随着智能机器人应用领域的日益扩大，人们期望智能机器人能在更多的领域为人类服务，代替人类完成更多更复杂的工作。

Development Trend of Intelligent Robots
智能机器人发展趋势

1. Brain-Like Intelligent Control

With supports of rich sensor data, robots can perform well in effective data classification and analysis, and extract reliable data in the future. The intelligent robots will be made with strong learning abilities, which can be divided into supervised learning and unsupervised learning.

Unsupervised learning mainly concentrates on constructing complicated neural network and deep learning. Deep learning mainly simulates the learning process of the human brain by using neural network. Robot can extract features during the training process and map relationship among robot motion, sensor information and tasks, thus helping robot finishing tasks without human interference, such as opening and closing doors, and grasping items(as shown in Figure 15-15).

Figure 15-15 Robot Training of Object Grasping   Video 59

2. Robots with Soft Structure

Traditional robots mainly have a rigid body with complicated structure, limited flexibility and poor safety and adaptation abilities. With the development of 3D printing technology and new intelligent materials, flexible robots made of soft or flexible materials are developed. Figure 15-16 shows examples of soft grippers.

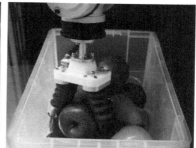

Figure 15-16 Example of Soft Grippers

3. Robot Based on Cloud Computing and Big Data

Cloud enabled robot is the combination of cloud computing and robotics. Cloud computing is a computing mode based on the Internet. The shareable software and hardware resources and information can be offered to network terminals according to demands. As the network terminal, robot does not need to store all data or own ultra-strong computing ability, but to link to related servers to take necessary information. An illustration of cloud enabled robot is shown in Figure 15-17.

Cloud enabled robot not only can upload complicated computing tasks to the cloud end, but also can receive massive data and share information and skills. With stronger abilities of storage, computing and learning, cloud enabled robots make resource sharing among robots more convenient.

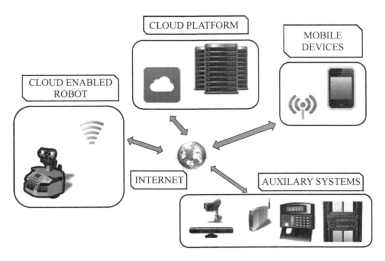

Figure 15-17　Cloud Enabled Robots

Vocabulary　词汇

1. concentrate   ['kɒnsntreɪt]   v. 使……集中（或集合、聚集）
2. complicated   ['kɒmplɪkeɪtɪd]   adj. 复杂的；难懂的
3. massive   ['mæsɪv]   adj. 巨大的；大量的
4. grasp   [grɑːsp]   v. 抓紧；抓牢；理解
5. flexibility   [ˌfleksə'bɪləti]   n. 柔韧性；灵活性；弹性；适应性
6. resource   [rɪ'sɔːs]   n. 资源
7. arouse   [ə'raʊz]   v. 激起；激发

Notes　注释

1. brain-like intelligent control　　　类脑智能控制
2. supervised learning　　　　　　　　监督学习
3. unsupervised learning　　　　　　　无监督学习
4. deep learning　　　　　　　　　　　深度学习
5. data classification　　　　　　　　数据分类
6. training process　　　　　　　　　 训练过程
7. rigid body　　　　　　　　　　　　刚体
8. soft gripper　　　　　　　　　　　软夹持器
9. intelligent material　　　　　　　智能材料
10. flexible robot　　　　　　　　　　柔性机器人

11. cloud enabled robot　　　　　　　　云机器人
12. network terminal　　　　　　　　　网络终端

Reference Translation　参考译文

1. 类脑智能控制

在丰富的传感器数据支持下，机器人能够很好地进行有效的数据分类和分析，并在未来提取可靠的有效数据。机器人具有较强的学习能力，可分为有监督学习和无监督学习。

无监督学习主要集中在复杂神经网络的构建和深度学习方面。深度学习主要利用神经网络模拟人脑的特征学习过程。机器人可以在训练过程中提取特征，绘制机器人运动、传感器信息和任务之间的关系曲线，从而帮助机器人在不受人为干扰的情况下完成开门、关门、抓物等任务，如图 15-15 所示。

2. 软结构机器人

传统机器人主要是结构复杂、柔韧性有限、安全性和适应能力差的刚性体。随着 3D 打印技术和新型智能材料的发展，由软材料或柔性材料制造的柔性机器人得到了发展。图 15-16 展示了软夹持器的例子。

3. 基于云计算和大数据的机器人

云机器人是云计算和机器人技术的结合。云计算是一种基于互联网的计算模式。可根据需要向网络终端提供可共享的软硬件资源和信息。作为网络终端，机器人不需要存储所有的数据或拥有超强的计算能力，而需要链接到相关的服务器上获取必要的信息。云机器人的示意图如图 15-17 所示。

云机器人不仅可以将复杂的计算任务上传到云端，还可以接收海量数据，共享信息和技能。它具有更强的存储、计算和学习能力，使机器人之间的资源共享更加方便。

先进制造业学习平台

先进制造业职业技能学习平台
工业机器人教育网（www.irobot-edu.com）

先进制造业互动教学平台
教学APP

一键下载
收入口袋

专业的教育平台	先进制造业垂直领域在线教育平台
更轻的学习方式	随时随地、无门槛实时线上学习
全维度学习体验	理论加实操，线上线下无缝对接
更快的成长路径	与百万工程师在线一起学习交流

领取专享积分

下载"教学APP"，进入"学问"—"圈子"，
晒出您与本书的合影或学习心得，即可领取超额积分。

先进制造业人才培养丛书书目

教学课件下载步骤

步骤一
登录"工业机器人教育网"
www.irobot-edu.com，菜单栏单击【学院】

步骤二
单击菜单栏【在线学堂】下方找到您需要的课程

步骤三
课程内视频下方单击【课件下载】

咨询与反馈

尊敬的读者：

  感谢您选用我们的教材！

  本书有丰富的配套教学资源，凡使用本书作为教材的教师可咨询有关实训装备事宜。在使用过程中，如有任何疑问或建议，可通过邮件（zhangmwen@126.com）或扫描右侧二维码，在线提交咨询信息，反馈建议或索取数字资源。

全国服务热线：400-6688-955

（教学资源建议反馈表）